ON THE LIVING CELL
A Theoretical Exploration of Biology

Baocheng Pan

细胞论——生物学的理论探讨

潘宝成 著

·广州·

版权所有　翻印必究

图书在版编目（CIP）数据

细胞论：生物学的理论探讨 = On the Living Cell：A Theoretical Exploration of Biology/潘宝成著. —广州：中山大学出版社，2018.3
ISBN 978 - 7 - 306 - 06310 - 6

Ⅰ.①细… Ⅱ.①潘… Ⅲ.①细胞生物学—研究 Ⅳ.①Q2

中国版本图书馆 CIP 数据核字（2018）第 044305 号

XIBAOLUN SHENGWUXUE DE LILUN TANTAO

出 版 人：徐　劲
策划编辑：林彩云
责任编辑：林彩云
封面设计：林绵华
责任校对：廖丽玲
责任技编：何雅涛
出版发行：中山大学出版社
电　　话：编辑部 020 - 84110771，84113349，84111997，84110779
　　　　　发行部 020 - 84111998，84111981，84111160
地　　址：广州市新港西路 135 号
邮　　编：510275　传　　真：020 - 84036565
网　　址：http://www.zsup.com.cn　E-mail：zdcbs@ mail.sysu.edu.cn
印 刷 者：广东省农垦总局印刷厂
规　　格：787mm×1092mm　1/16　11.5 印张　258 千字
版次印次：2018 年 3 月第 1 版　2018 年 3 月第 1 次印刷
定　　价：45.00 元

如发现本书因印装质量影响阅读，请与出版社发行部联系调换

> I think, therefore I am.
>
> ——René Descartes

PREFACE

Life is mysterious! We human beings have been searching for the mystery since the time of Aristotle and even now we are still striving for it. There are not any other scientific disciplines like biology attracting so many researchers and investigators since last decades. Since the 1980s, biology has made many achievements, especially in the field of molecular biology. More recently, the research findings in the field of stem cell biology provided a lot of new insights into the properties of the cell and improved our understanding of the essentials of the cell. Indeed, the experimental results in biology have been accumulated enormously. I am confident that it is high time for us to investigate these experimental results from the theoretical viewpoints and to explore theoretical framework of biology.

The living system and the non-living system are the two fundamentally different systems in the natural world. These two kinds of systems display essentially different features. While philosophy and methodology for the non-living systems are quite sophisticated, we are now still basically at the experimental stage of the study of the living system and most of the current research mainly concentrates on experimental investigation. Undoubtedly, experimental biology has been advancing by leaps and bounds in recent decades. Comparatively, however, theoretical biology has made little progress. Indeed, investigation of the living system still remains at the stage of experimentation, which is the preliminary stage of scientific research. Theoretical biology is a new land with hope and fruits and needs to be explored and cultivated. Even though it seems amazing and attractive, theoretical exploration of biology is very challenging and even risky. In this book I am willing to take such a risk.

On the Living Cell—A Theoretical Exploration of Biology

The first and maybe the most difficult hurdle for theoretical study of biology comes from the traditional scientific concepts rooted in the philosophy of sciences, which may be more accurately termed philosophy of physical sciences. We have been so accustomed to these concepts which come directly from physical sciences that we take it for granted that they are also applicable to biological sciences and that furthermore they can explain all the phenomena and activities of biological systems. First of all, therefore, we need to investigate the basic concepts from the philosophical perspectives to determine whether or not they are actually valid for biological sciences.

Biology can be classified into two branches, functional biology and evolutionary biology. Functional biology studies structures, activities and interactions and uses experimentation as its research method. Evolutionary biology, on the other hand, investigates historical aspects and uses comparative method. Functional biology uses the similar concepts and methods as physical sciences but evolutionary biology introduces new concepts and uses new methods that are different from physical sciences. Functional biology studies "how" questions and evolutionary biology investigates "why" questions (Mayr, 1997; 2004). The research in this book falls in the category of functional biology, namely, investigation of the "how" questions in biology. Take teleonomy of the living system for an example. The book will not investigate the source from which the system has obtained this property, which is the work of evolutionary biology. The book only studies how this property affects the behaviors and characteristics of the system. Even so, I still want to emphasize that this property should not be classified into the category of vital force or any other supernatural forces. Instead, the property presumably results from the total interactions between all the components of the system, namely, the emergent property of the system when the parts form the whole.

The book mostly studies the basic unit of the living system, namely, the cell. The cell is a living system with many biological components and inorganic elements. It is not a simple machine made up of chemicals. It has its own power and its own intention. If it has to be regarded as a machine, then it must be a

PREFACE

machine with at least an engine because it has its own driving force to fulfill its tasks without any help, signals or directives from the environment.

The book presents one axiom and four basic hypotheses for theoretical biology on cellular level based on my understanding of current concepts and knowledge of biology, with an attempt to use them to describe the principles and mechanisms of biological processes in the living cell. It has six chapters. In Chapter One, the fundamental issues in theoretical biology are discussed, including the limitation of traditional philosophy of sciences, the autonomy of biology, the dual attributes of the living system, and the systems that are most suitable for theoretical study. In Chapter Two, the axiom of survival is proposed and discussion of purposefulness in biology is given mostly based on philosophical considerations. In Chapter Three, the hypothesis of orderliness of the cell is suggested and its application to senescence is discussed. In Chapter Four, the hypothesis of mitosis and cell growth and the hypothesis of meiosis and cellular genetic diversity are proposed and the related corollaries are given. Also the economic principle of cell division is discussed. In Chapter Five, the principle of cell differentiation is hypothesized and the transitions between different types of cells are discussed. In the last chapter, the similarity between the machine code of artificial intelligence and the genome of the cell is investigated. Based on such similarity, a conjecture is presented about the "Final Rules" in theoretical biology.

At present, theoretical biology is still in its preliminary stage. Generally speaking, most of the current works on theoretical biology, in my opinion, should be essentially classified as the works of computational biology. Basically, these researches started with some mathematical models obtained by simplifying some processes of biological systems, and performed calculations using analytical or computational methods of mathematics, and then compared and discussed these results with the experimental results obtained in related biological processes. More importantly, these works usually simplified the living system into physical or chemical system and ignored the fundamental differences between the living system and the non-living system. However, I think that the main goal of theoretical biology is not just to do such model computation, but to start from the autonomy of biology, and explore the basic concepts and characteristic quantities unique to biology, and establish its own theoretical system.

Writing the book was indeed a tremendously difficult and risky work. There were so many philosophical and scientific hurdles to overcome and the direction and paths for the establishment of theoretical biology were still unclear. In the past seven years, I consulted a lot of literature, including some classical works by pioneers in the fields of biological sciences. There were myriad experimental results and philosophical discussions about the similar topics. Some of them were equivocal and even contradictory to each other. What was worse, there were so few theoretical analyses and concepts available for reference. Indeed, theoretical biology was a brand new field. There was nothing to follow. Therefore, exploration in this field needed not only wisdom but also courage because many years of hard work might end up with nothing.

What was the starting point for the exploration of theoretical biology? After deep consideration for a long time, I thought that philosophy of sciences might be the best candidate because it provided the most fundamental perspective of the nature and could free us of the restriction of the knowledge and concepts of specific disciplines. Once the starting point had been set, I had to keep working hard. The odyssey was full of ups and downs. Sometimes I felt the endeavor to explore the theory of biology as if looking for a needle in a haystack. The truth of biological world looked like a mirage on the Promised Land. It was so attractive but so mysterious at the same time. Sometimes the truth seemed so close to me that it was near my fingertips and sometimes it looked so far away from me that I did not know how to march forward. It was my belief in the simplicity of the Nature and the existence of the theory that kept me going on and on. I was sure that the more I did the closer I would be to the truth.

Based on consideration of the fundamental differences between the living system and the non-living system, the book is intended to explore the principles that govern the processes of living systems. These principles should be

PREFACE

independent of the physical sciences and peculiar to biological sciences. Obviously the theory in this book is preliminary and I do not expect all the statements and discussions in the book are correct. Instead, I would be fully satisfied if my work can serve as a starting point for the theoretical exploration of the principles in the biological world. Even though I know that the theory is immature I still want to share with you because exploration of theoretical biology is a gigantic endeavor and the establishment of theoretical biology relies on participation and cooperation of scientists from various disciplines and different fields. I am confident that when many people come together to investigate theoretical biology we can eventually unravel the mysteries of life. I hope that the immature but interesting concepts and ideas in the book can stimulate your interest in the fundamental issues of biology and philosophy and inspire your enthusiasm to search for the laws and principles of the biological world. I expect that the field of theoretical biology will be more and more promising in the near future.

I remember an interesting story. The main script is as follows: In a dark night, a passerby saw a person looking for something under streetlight and asked: "What are you looking for?" The person answered: "I was looking for a key." The passerby asked: "Where did you lose your key?" The person answered: "I did not know." The passerby asked further: "Why are you looking for it here?" The person answered: "Because there is light here and it is the only place that I can see." This simple story conveyed an important message that if we overemphasize the guidance of the existing concepts and theories, we may get trapped into the cage of the old concepts and theories.

In writing this book, I received a lot of support and help from Professor Tinghuai Wang at Xinhua College of Sun Yat-sen University. He provided me with a stable environment in the campus and gave me continuous encouragement in my work so that I could concentrate myself on the final stage of writing the book. Here, I show my greatest gratitude to him for his great support.

During the whole course of writing the book. I had a lot of discussions and consultations with my colleagues and friends: especially Yu Peng, Shiney

Zhang, Donna Lai, Xiangjun Chen, Li Feng, Yong Xiong, Dehua Pei and Xiaoyu Liu. They provided many constructive suggestions and opinions, which improved the book very much. Without their help and support, the book would not be finished. I hope that the book could serve as a thankful gift to all of them for their help and encouragement.

B. P.

Xinhua College

Sun Yat-sen University

Guangzhou, China

August 22, 2017

> 我思故我在。
>
> ——雷内·笛卡尔（*René Descartes*）

前　　言

　　生命是神秘的！从亚里士多德时代起，我们人类就一直努力不懈地探索生命的奥秘，甚至到现在我们还在为之奋斗。近几十年来，没有任何一门科学学科像生物学那样吸引着无数研究人员。从20世纪80年代至今，生物学取得了长足的进步，分子生物学领域表现尤为突出。最近，干细胞生物学领域的研究进展为细胞的属性提供了不少新见解，加深了我们对细胞本质的理解。确实，生物学的实验结果已经累积得相当多了。我深信，现在是我们从理论的角度来研究这些实验结果、探讨生物学理论框架的时候了。

　　生命系统和非生命系统是自然界中两个根本不同的系统。这两类系统显示出截然不同的特性。虽然我们对非生命系统的研究在理念和方法上已经相当严谨、成熟，但对生命系统的研究我们基本上还处于实验阶段，目前大部分工作主要还集中在实验研究上。毫无疑问，实验生物学在近几十年来发展神速。相对而言，理论生物学研究进展不大。确实，对生命系统的研究仍然停留在实验这样一个科学研究的初级阶段。理论生物学是充满希望和成果的"新大陆"，有待我们去开发和耕耘。它虽然看上去似乎精彩迷人，但事实上生物学的理论探索极具挑战性，甚至是要冒险的。在这本书里我愿意冒这样的风险。

　　生物学理论研究的第一个障碍或许也是最难以克服的障碍来自于根植于科学哲学的传统的科学概念，而这种科学哲学更准确地说应该是物理科学的哲学。然而，我们已经太过习惯于这些来自物理科学的概念，以至于理所当然地认为它们同样适用于生物学，并且能够用来解释生物系统所有的现象和行为。因此，首先我们需要从哲学的角度来考查这些基本的科学概念，以确定它们是否对生物学依然有效。

　　生物学可分为功能生物学和进化生物学两大分支。功能生物学研究结构、活动和相互作用，并以实验为研究方法。而进化生物学研究历史特性和使用比较方法。功能生物学使用与物理科学类似的概念和方法，但进化生物学引入新的概念，使用不同于物理科学的新方法。功能生物学研究"如何"的问题，而进化生物学研究"为什么"的问题（Mayr，1997，2004）。本书的研究属于功能生物学范畴，即生物学中"如何"问题的研究。以生命系统的目的性为例。这本书不会考查系统获得这个属性的来源，这是进化生物学的工作。这本书只研究这个属性如何影响系统的行为和特性。即便如此，我仍然想强调，这种属性不应归入生命力或任何其他超自然力量的范畴，相反，这种属性可能是由系统各组成部分之间的总的相互作用引起的，也就是说，当各部分组成整体时系统所出现的新质。

这本书主要研究生命系统的基本单元——细胞。细胞是一个具有多种生物成分和无机元素的生命系统。它不是一个由化学物质构成的简单机器。它有自己的动力和意图。如果它必须被视为一台机器，那么它至少是一台有发动机的机器，因为它有自己的驱动力去完成任务，而不需要任何帮助或者来自环境的信号或指令。

基于我对现有的生物学概念和知识的理解，本书在细胞水平上提出了理论生物学的一个公理和四个基本假设，试图用它们来描述活细胞中生物过程的原理和机制。全书共有六章。第一章讨论了理论生物学的基本问题，包括传统科学哲学的局限性、生物学的自主性、生命系统的属性二象性以及最适合理论研究的系统。第二章提出了生存公理，并且主要从哲学角度来讨论生物学的目的性。第三章提出细胞的有序性假设，并讨论其在衰老中的应用。第四章提出了有丝分裂和细胞生长假设以及减数分裂和细胞遗传多样性假设，并且给出了相关的推论。此外，还讨论了细胞分裂的经济原则。第五章提出了细胞分化原理假设，并讨论了不同类型细胞之间的转变。最后一章研究了人工智能机器代码与细胞基因组的相似性。基于这种相似性，我提出了关于理论生物学的"终极法则"猜想。

当前，理论生物学仍处于初级阶段。综观目前大部分理论生物学的研究工作，在我看来，本质上应该属于计算生物学的工作。这些研究基本上是以通过简化生物系统的某些过程而获取的一些数学模型为出发点，然后运用数学分析或计算机方法进行运算，再比较和讨论这些结果和相关生物过程的实验结果之间的关联。更重要的是，这些研究将生命系统简化成物理或化学系统来进行研究，而忽视了生命系统和非生命系统之间的本质区别。然而，我认为理论生物学的主要目的并不应该只是做这样一些模型计算，而应该是从生物学的自主性出发，发掘生物学本身特有的基本概念和特征量，建立其自身的理论体系。

这本书的写作确实是一项极其困难和冒险的工作，要克服许多哲学上和科学上的障碍，而且建立理论生物学的方向和途径还不明确。在过去的 7 年中，我查阅了大量的文献，包括一些生物科学领域先驱者的经典著作。关于类似课题的实验结果和哲学讨论都有很多，其中有些是模棱两可的，有些甚至是互相矛盾的。更糟糕的是，很少有理论分析和概念可供参考。确实，理论生物学是一个崭新的领域，没有什么可跟随的。因此，在这一领域的探索不仅需要智慧，更需要勇气，因为多年的辛勤工作最终可能是一无所获。

探索理论生物学的出发点是什么？经过长时间的深入思考，我认为科学哲学可能是最好的选择，因为它提供了最基本的自然观，可以使我们摆脱特定学科知识和概念的限制。出发点一旦确定，我就要努力工作了。探索的征程充满了跌宕起伏，有时我觉得探索生物学理论就像大海捞针一样。生物世界的真理看起来就像是乐土上的海市蜃楼，它很吸引人，但同时又很神秘。有时真理似乎离我很近，它就在我的指尖，有时它离我很远，我不知道如何向前走。正是对自然简单性和理论存在性的笃信让我不断地前行。我深信做得越多，我就越接近真理。

考虑到生命系统和非生命系统的本质区别，本书旨在探索支配生命系统过程的原理。这些原理应该独立于物理科学，而且是生物科学所特有的。显然，这本书中的理论是初步的，我并不期望书中所有的陈述和讨论都是正确的。相反，如果我的工作能够作为对生物世界中基本原理的理论探索的起点，我将深感欣慰。虽然我知道理论还不成熟，但我还是想和大家分享，因为对理论生物学的探索是一个巨大工程，并且理论生物学的建立有赖于各个学科和领域的科学家的参与和合作。我相信，当许多人聚集在一起研究理论生物学时，我们最终可以揭示生命的奥秘。我希望书中那些不成熟但有趣的概念和想法能够引起您对生物学和哲学基本问题的兴趣，能激发您对探索生命世界的定律和原理的热情。我期待理论生物学领域在不久的将来会有越来越广阔的发展前景。

我记得一个有趣的故事。故事的大致情节如下：在一个漆黑的夜晚，一个过路人看到一个人在路灯下找东西，便问道："你在找什么？"那个人回答说："我在找钥匙。"过路人问："钥匙是在哪里丢的？"那个人回答：

"我不知道。"路人问:"那你为什么要在这里找呢?"那人回答说:"因为这里有灯光,这是我唯一能看到的地方。"这个简单的故事传递了一个很重要的启示,如果我们过分强调现有概念和理论的指导作用,我们有可能会被困在旧概念和旧理论的笼子里。

在这本书的写作过程中,我得到了中山大学新华学院王庭槐教授的大力支持和帮助。他给予我很多鼓励并在校园里为我提供了安定的工作环境,使我得以专注于这本书的后期写作。在此表示衷心的感谢。

在这本书的整个写作过程中,我与同事和朋友们进行了大量的讨论和磋商,特别是 Yu Peng, Shiney Zhang, Donna Lai, Xiangjun Chen, Li Feng, Yong Xiong, Dehua Pei 和 Xiaoyu Liu。他们提出了许多建设性的建议和意见,大大改进了这本书的内容。没有他们的帮助和支持,这本书就不会完成。我希望这本书能成为感谢他们的帮助和鼓励的礼物。

潘宝成
2017 年 8 月 22 日
于中国广州中山大学新华学院

CONTENTS

Chapter One Fundamental Issues in Theoretical Study of Biology ······ 1
  1.1 Background ·· 1
  1.2 The Autonomy of Biology ·· 5
  1.3 The Dual Attributes of the Living System ···························· 8
  1.4 The Cell as the Target for the Research of Theoretical Biology
       ·· 13

Chapter Two Survival and Purposefulness in Biology ···················· 17
  2.1 Background ·· 17
  2.2 The Axiom of Survival ·· 17
  2.3 The Relationship of the Axiom of Survival and Cell Death ······ 19
  2.4 Purposefulness in Biology ·· 22
  2.5 The Significance of the Axiom of Survival ························ 30
  2.6 A Definition of Life ·· 31

Chapter Three Orderliness and Senescence of the Cell ···················· 33
  3.1 Background ·· 33
  3.2 Orderliness of the Living System ·· 34
  3.3 Hypothesis of Orderliness of the Cell ································ 40
  3.4 Senescence of the Cell ·· 43
  3.5 Orderliness and Senescence ·· 47

Chapter Four The Principle of Cell Growth and Cell Division ·········· 51
  4.1 Background ·· 51
  4.2 Relationship between Cell Division and Axiom of Survival ······ 52
  4.3 Hypothesis of Mitosis and Cell Growth ···························· 54
  4.4 Hypothesis of Meiosis and Cellular Genetic Diversity ········ 60
  4.5 The Economic Principle of Cell Division ·························· 63

Chapter Five The Principle of Cell Differentiation ……………… 66
   5.1   Background ……………………………………………………… 66
   5.2   Genome as the Identification of the Somatic Cell ………… 66
   5.3   The Differentiation Potency and Differentiation Potential of the Somatic Cell ………………………………………………… 67
   5.4   Hypothesis of Cell Differentiation ……………………………… 70
   5.5   Relationship of Differentiationality and Functionality ……… 74
   5.6   The Tree Structure of the Somatic Cells in Multicellular Organisms ……………………………………………………………… 76
   5.7   Transition between Different Differentiation Potency States …… 78

Chapter Six A Conjecture of the "Final Rules" of Theoretical Biology ……………………………………………………………… 80
   6.1   Background ……………………………………………………… 80
   6.2   Artificial Intelligence ……………………………………………… 82
   6.3   Some Features of the Genome of the Living Cell ………… 85
   6.4   The "Final Rules" of Theoretical Biology ……………………… 88
   6.5   Implications of the "Final Rules" ……………………………… 93

中文翻译 ………………………………………………………………… 96
第一章　生物学理论研究的基本问题 ………………………………… 96
   1.1   背景知识 ………………………………………………………… 96
   1.2   生物学的自主性 ………………………………………………… 99
   1.3   生命系统的属性二象性 ………………………………………… 101
   1.4   细胞作为理论生物学研究的对象 ……………………………… 104

第二章　生存与生物学的目的性 ……………………………………… 107
   2.1   背景知识 ………………………………………………………… 107
   2.2   生存公理 ………………………………………………………… 107
   2.3   生存公理与细胞死亡的关系 …………………………………… 108
   2.4   生物学的目的性 ………………………………………………… 111
   2.5   生存公理的意义 ………………………………………………… 116

2.6 生命的定义 ··· 117

第三章 细胞的有序性及其衰老 ·· 119
  3.1 背景知识 ··· 119
  3.2 生命系统的有序性 ··· 119
  3.3 细胞的有序性假设 ··· 123
  3.4 细胞的衰老 ··· 125
  3.5 有序性与衰老 ··· 128

第四章 细胞生长和细胞分裂原理 ·· 131
  4.1 背景知识 ··· 131
  4.2 细胞分裂与生存公理的关系 ····································· 132
  4.3 有丝分裂和细胞生长假设 ·· 133
  4.4 减数分裂和细胞遗传多样性假设 ····························· 137
  4.5 细胞分裂的经济原理 ··· 139

第五章 细胞分化原理 ·· 141
  5.1 背景知识 ··· 141
  5.2 基因组作为体细胞的身份标识 ································ 141
  5.3 体细胞的分化潜能与分化势能 ································ 142
  5.4 细胞分化假设 ··· 144
  5.5 分化性与功能性的关系 ··· 146
  5.6 多细胞生物体中体细胞的树状结构 ························ 148
  5.7 不同分化潜能态之间的转换 ···································· 149

第六章 理论生物学的"终极法则"猜想 ·· 151
  6.1 背景知识 ··· 151
  6.2 人工智能 ··· 152
  6.3 活细胞基因组的一些特征 ·· 155
  6.4 理论生物学的"终极法则" ···································· 157
  6.5 "终极法则"的含意 ··· 160

References ·· 163

All the answers to biology are eventually sought in the cells because all organisms are or once were a cell.

——EB Wilson

Chapter One
Fundamental Issues in Theoretical Study of Biology

1.1 Background

What is life? The answer to this seemingly simple question has been being searched for by many philosophers and scientists since the time of Aristotle but until now we still do not have a clear and satisfactory answer. Erwin Schrödinger asked this question in the 1940s and tried to answer it within the framework of physics. He considered the chromosome as an aperiodic solid in which every group of atoms plays an individual role and proposed some concepts such as negative entropy flow to sustain the living system (Schrödinger, 1944). Ernst Mayr studied the meaning of life from the philosophical prospective. He proposed that all biological processes are controlled not only by natural laws but also by genetic programs (Mayr, 1997; 2004). In other words, all living organisms obey two causalities. One is the natural laws that, together with chance, control completely everything that happens in the world of the exact sciences. The other causality consists of the genetic programs that characterize the living world uniquely. The dualism of modern biology is basically physicochemical, and it arises from the fact that organisms possess both a genotype and a phenotype (Mayr, 1997; 2004). Stuart Kauffman studied life from the viewpoint of holism and believed that life is the emergent phenomenon which arises when the molecular diversity increases beyond some threshold values of complexity (Kauffman, 1995). Therefore life is a collective property of systems of interacting molecules. Addy Pross investigated living systems from the basis of systems chemistry. By introducing the concept of dynamic kinetic stability, he claimed that biology can be integrated into chemistry (Pross, 2012). Marcello Barbieri listed more than 60 typical definitions of life from scientists working in different fields in his book entitled "The Organic Codes" (Barbieri, 2003). Many of these definitions are so different that it is difficult for us to imagine that they describe the same thing.

Indeed, the world we are living in is so diverse and complex that scientists in different fields or with different background have acquired different impressions and understanding of the world. The definition of life discussed above is a good example. However, the Nature is simple. I think that if a scientific theory is not simple, that is because the scientific research has not revealed the essence of the Nature. The theoretical exploration of biology may start with classification of the systems in the world. Everything in the world can be classified into either living system or non-living system. To further study what life is and what principles that govern living systems, first we need to clarify the definition of the living system and the non-living system. Usually living systems are defined as any systems that have the characteristics of replication and metabolism. Here and hereafter in the book, the living system is defined for any systems that have life, and the non-living system is defined for any systems that do not have life, including those biological systems that are only composed of biological macromolecules but do not have life. I will give a definition of life in Chapter 2.

Now let us study different behaviors between the living systems and the non-living systems. Take a look at a fish and a piece of woody log in a river (Figure 1.1). As we all know, this piece of log flows downstream along with water current. The river terrain may be very complicated but, in principle, we can describe the motion trajectory of this log and can predict its locations at any time in the future according to the principles and laws of physical sciences. However, we cannot do that at all for the fish. We do not know how and where the fish will move in the next moment; we cannot predict the motion trajectory of the fish with any principles or laws of physical sciences. Similarly, we can predict that a ball always rolls down a slope but we do not know how and where an ant on the same slope will move. There are ample such examples around us. These examples reveal the fundamentally different behaviors of the living system and the non-living system and also show the limitations of the physical sciences in describing the behaviors of the living organisms.

Figure 1.1　Live Fish and a Woody Log

The different behaviors of live fish and a woody log indicates that the living system and the non-living system follow different principles in nature (drawn by Yu Peng).

In addition to behavior unpredictability, another obvious difference between the living system and the non-living system is the behavior non-repeatability. Physics and philosophy of science consider that the objectivity of an event is closely associated with the repeatability of its observation. The events observed in experiment must be repeatable before they can be accepted in science. However, for the living system, objectivity and repeatability are not always consistent. For example, in the previously mentioned slope, the same ant in different time may have different motion trajectories, sometimes going up the slope and sometimes going down the slope. Comparatively, the same ball in different time will have the same motion trajectory, always going down the slope. This is another fundamental difference between the living system and the non-living system. Obviously, we cannot deny the objective existence of the motion trajectory of the ant because of its non-repeatability of the motion trajectory.

On the Living Cell—A Theoretical Exploration of Biology

The purpose of science is to find some laws to describe the behaviors and characteristics of the entities in the world. Scientific research started with non-living systems because they are much easier to understand and simpler to study than living systems. Beginning from Galileo to Newton and to the establishment of the theory of relativity and quantum mechanics, investigation of the non-living systems have achieved tremendous success, which made us take it for granted that everything in the world, including the living system, obeys the same laws of physical sciences and that the differences between the living system and the non-living system do not play an important role in our understanding of the world. Indeed, the history of scientific development showed that physical sciences dominated over every field of science. Philosophy of science and its basic ideas and concepts are established and developed mostly based on the experimental results and theoretical analyses of physical sciences. Since the birth of classical mechanics, our understanding of the universe and life has been imprinted with the marks of physical sciences. We are so accustomed to these ideas and concepts that we instinctively accept anything that is consistent with them and reject anything that is inconsistent with them. Traditional philosophy of sciences took physics as the norm or standard paradigm for all scientific disciplines and never treated biology as an independent discipline of science. Reductionists believed that all the phenomena in the biological world can be eventually explained by the principles of physical sciences.

Stephen Hawking and Leonard Mlodinow pointed out that scientific philosophy has lagged far behind the development of modern sciences. In the book "The Grand Design", they stated that "Traditionally these (such as 'How does the universe behave?' and 'What is the nature of reality?') are questions for philosophy, but philosophy is dead. Philosophy has not kept up with modern development in science, particularly physics." (Hawking & Mlodinow, 2010) In my opinion, the situation in biology is much more severe than that in physics.

Since the 1980s biology has been advancing by leaps and bounds, especially in the fields of molecular biology and stem cell biology. As pointed out by Darwin that the achievements in biology would make philosophy a new prosperity. Indeed, the rapid development in biology resulted in a new philosophy of science, the philosophy of biology. The new philosophy emphasized that biology should get rid of the bondage of traditional philosophy of science and establish its own theories, concepts and structures. One of the most fundamental questions of philosophy of biology is the status of biology in natural sciences. More specifically, is biology an autonomous discipline of science that is independent of the framework of physical sciences or is only a branch of physical sciences within the framework? This question is of crucial importance for theoretical biology because it determines not only the framework of theoretical biology but also the direction and path of the development of theoretical biology. Among many of the active advocates of the philosophy of biology, Mayr is one of the most important figures. He studied biology especially evolutionary biology intensively and analyzed the roles of biology in science and concluded that biology is an autonomous discipline in science (Mayr, 1996). He said that "More and more clearly I began to see that biology was a quite different kind of science from the physical sciences; it differed fundamentally in its subject matter, its history, its methods, and its philosophy." (Mayr, 1988)

The living system is enormously complex and the phenomena we observed are usually interwoven with the intrinsic properties of the system and its responses to the interventions of the environment. In order to establish a theoretical framework of biology, I think that, first of all, we need to identify the intrinsic properties and the stimulated responses of the living system from its behaviors and activities in biological processes. Only after distinguishing the internal and external factors that determine the behaviors and activities of the living system would it be possible for us to explore the theory of biology in a correct way. The book will mostly study the intrinsic properties of the living system.

1.2 The Autonomy of Biology

There are two opposite schools for the status of biology in science: reductionism and autonomy. The reductionists consider that the living system and the non-living system are made up of the same chemical elements and therefore in principle the living system should obey the same principles and laws as those for the non-living system. They believe that biology is not independent of physics and the research of molecular biology will eventually reduce the whole biology to physics and hence biology will be only one branch of physical sciences. Reductionists use some strategies to refute the autonomy of biology. For instance, they classify biology into two parts, the first part can be reduced to physical sciences and obeys the universal laws while the second part cannot be reduced to physical sciences and is thus regarded as non-scientific (Mayr, 1996). Therefore, the reductionists claimed that there is no need to acknowledge autonomy of biology. With tremendous development in biology since the 1980s, scientists began to realize that physics and its related classical philosophical concepts are not sufficient to explain the phenomena observed in biological sciences. The development of biology led to the establishment of another school of philosophy of biology, namely, autonomy of biology, which is opposite to reductionism. The proponents of autonomy consider that the research objects of biology, conceptual structures and methodology in biology are totally different from those of physical sciences. They believe that biology cannot be reduced to physical sciences and the principles of life should be established on a new framework that is independent of physical sciences.

The reason that the reductionists consider biology non-autonomous was originated from their philosophical belief of reductionism. Reductionists believe that the explanation of a system can be achieved in principle by studying its smallest components and the knowledge of the system will be obtained from its decomposed smaller components (Mayr, 2004). Certainly, a system can be understood in principle by studying its smaller components. To this end, however, some conditions must be met. For instance, we need to know the interaction mechanism between the smaller parts, by which the system is formed. Let us do a thought experiment for the living system. Imagine we decompose a living system into smaller components which belong to non-living system. Even though we understand completely these smaller components, we still cannot understand the intrinsic properties of the living system as a whole or obtain the knowledge of the living system as a whole from the smaller components because we still do not know the transformation mechanism from a non-living system into a living system. In fact, when we decompose a living system into smaller non-living systems, we have already "killed" the living system. In this situation, the living system loses its characteristics and information of life, which cannot be recombined from the non-living components. Therefore, reductionism is confronted with great challenges in philosophy of biology.

It is high time for us to reconsider the applicability of the laws and principles of physical sciences to the living system. Considering the fundamental characteristics of the living system, the laws and principles of physical sciences, such as the principles of energy minimization and entropy maximization, are neither direct nor effective in describing the features and behaviors of the living system. For instance, the most favorable state for the non-living system is the most stable state, such as the state with the lowest energy for dynamic systems or with the maximal entropy for the thermodynamic systems. Obviously, the most stable state for the non-living system is not at all among the favorable states for the living system because the state with lowest energy or maximal entropy means the end of life to the living system and therefore should be avoided. It is clear that the living system and the non-living system are endowed with essentially different characteristics and therefore should follow different laws and principles in their own particular processes. The states that we are interested in for the living system are not the most stable states but the states in which the living system has the best functions and activities. One of the purposes of biological sciences is to search conditions that optimize the functions and activities of the living system but not to search conditions that are used to reach states with lowest energy or maximal entropy. Therefore, the physical concepts and physical quantities such as stability, energy and entropy are not the best nor the most effective terms in describing the living system. Biology should have its own concepts and quantities in its theories. These concepts will come directly from the phenomena in the biological world and are independent of the physical world. These autonomous concepts are basic and original elements for the theory and represent the intrinsic features of life and therefore cannot be and should not be explained by physical sciences. We need a completely new set of theoretical framework to describe the living system.

1.3 The Dual Attributes of the Living System

Let us start our discussion with the same example we had before, a woody log and a fish in a river (Figure 1.1). We know that the fish will move in a trajectory that we cannot predict. How about a dead fish? We have the experience that it will move like a log in the river. Let us make a comparison between a live fish and a just dead fish. The chemical compositions of the two fish should be essentially the same but their behaviors are totally different, one follows the laws of physical sciences while the other does not. Obviously the only difference is that one has life while the other does not. Such difference solely results from the nature of life. It is the nature of life that makes the complete difference in behavior of the living system and the non-living system. In other words, life endows the living system with some particular characteristics that the non-living system does not possess. These characteristics disappear once life is gone from the living system. This simple example leads to a profound conclusion: the living system has an intrinsic characteristic of dual attributes: one is physicochemical and the other biological. The physicochemical attribute of the living system obeys the laws and principles of physical sciences while the biological attribute follows the laws and principles of autonomous biology.

What are the roles of these two different attributes play in the behavior of the living system? Let us study another example: a live fox falls down from a cliff. In this case, no matter what the fox thinks and what it wants to do, the gravitation of the earth plays the crucial role in its trajectory of motion. Combining with the discussion about Figure 1.1, we may conclude that the dual attributes of the living system are the two factors that determine the behavior and trajectory of the living system. Which attribute plays the dominant role in the behavior and trajectory of motion depends on the situation in which the living system is involved. By comparison, the non-living system only has physicochemical attribute and hence its behaviors are completely governed by the laws and principles of physics and chemistry. Therefore, the behaviors of the non-living system can be explained and predicted by physics and chemistry.

When the living system loses its life, it will become a non-living system and will only have physicochemical attribute.

The above discussion indicates that the living system has duality in attribute, one represents the characteristics in the inanimate world and follows the laws and principles of physical sciences and the other represents the characteristics in the animate world and follows the laws and principles that are independent of physical sciences. Therefore the biological attribute plays an important role in our understanding of life.

Then where does the biological attribute come from? First of all we need to examine the features in the structure of the living system and the non-living system. The great difference between the two lies in the high complexity in organizational structure of the living system, especially the hierarchical structure, which is due to the interaction between the organizations at a hierarchical level to form the organizations at a higher hierarchical level. The complexity of these structures endows the living system with particular features and capability, such as the response to external stimuli, the ability to grow, proliferate and differentiate (Mayr, 2004). Generally, emergent properties come into being when the parts form the whole. This phenomenon exists in both the living system and the non-living system. Holism states that "the whole is more than the sum of its parts", the emergent properties cannot be deduced from reconstruction of the knowledge of the component parts of the whole. For instance, aquosity of water molecule emerges from formation of hydrogen atom and oxygen atom but could not be deduced from the characteristics of hydrogen atom and oxygen atom (Mayr, 1997). Living organisms form a hierarchy of even more complex systems, from cells, organs to the whole organisms. The characteristics of an organization at each higher hierarchical level that emerge from its components at lower hierarchical level cannot be deduced from the knowledge of the components. Obviously, the emergent properties are even more tremendous and surprising when the parts belong to the non-living system and the whole is the living system, such as in the formation of the cell from its molecular components. Therefore, the emergent properties resulting from the formation of the living system from the non-living systems should be regarded as the intrinsic properties of the living systems and the source of biological attribute of the living system.

Considering the fact that the living system has dual attributes, namely, the physicochemical attribute and biological attribute, and that the biological attribute cannot be explained by physical sciences, we should not be surprised if some of the emergent properties of the living system cannot be explained by the principles and laws of physical sciences. In fact, it is these emergent properties that should be the essential part of biological study and make biology independent of physical sciences.

Mayr pointed out that if one could plot the domains of the physical and biological sciences on a map, one would find a considerable area of overlap. The overlapped area represents the common part shared by both physical and biological sciences while the non-overlapped area displays the irreconcilable differences between living beings and inert matter (Mayr, 1996). I think that the overlapped and non-overlapped domains, in some sense, confirm the existence of the dual attributes of the living system. Therefore, the attribute duality of the living system conforms to the knowledge structure of physical and biological sciences.

Mayr also proposed dual causation for biological processes of the living system. He stated that all living processes obey two causalities, one is the natural laws in the exact sciences while the other is the genetic programs that characterize the living world (Mayr, 1961; 2004). He stated that "Both worlds (the inanimate and the living world) obey the universal laws discovered and analyzed by the physical sciences, but living organisms obey also a second set of causes, the instructions from the genetic program. This second type of causation is nonexistent in the inanimate world." (Mayr, 1997) Mayr further explained that "their (living organisms') activities are governed by genetic programs containing historically acquired information, again something absent in inanimate world ... The dualism of modern biology is consistently physicochemical, and it arises from the fact that the organisms possess both genotype and phenotype. The genotype, consisting of nucleic acids, requires for its understanding evolutionary explanations. The phenotype, constructed on the basis of the information provided by the genotype, and consisting of proteins, lipids, and other macromolecules, requires functional (proximate) explanations for its understanding. Such duality is unknown in the inanimate world. Explanation of the genotype and of the phenotype require different kinds of theories." (Mayr, 1997)

It is clear from the above discussion that the dual causation proposed by Mayr is associated with both proximate and ultimate causes of the living system. One of the dual causation belongs to the category of functional biology in the form of phenotype while the other belongs to the category of evolutionary biology in the form of genotype. Both causations are physicochemical in nature. By comparison, the dual attributes of the living system I propose here is different from the dual causation proposed by Mayr. In my dual attributes of the living systems, one of the attributes is physicochemical and the other is biological. Both attributes only refer to the proximate causation of the living systems, namely within the category of functional biology. As discussed exclusion of, the biological attribute in the duality is not metaphysical or vitalistic; it is in nature the emergent property of the living system resulting from the interactions among its non-living components.

The concept of duality is not uncommon in science. For instance, the wave-particle duality is the characteristic of the particles in the microscopic world. The attribute duality of the living system may be one of the basic characteristics in the living world.

In fact, it is the biological attribute of the living system that makes biology autonomous. Unfortunately, most of the current research basically studies the physicochemical attribute instead of the biological attribute of the living system. There may be three main reasons: 1) unawareness of the existence of the biological attribute and regarding the physicochemical attribute is the only attribute of the living system; 2) ignorance of the importance of the biological attribute; 3) exclusion of the biological attribute from the scope of scientific research because of its "non-objective" characteristic.

The living system can be studied at different levels of hierarchical structures, from individual organism, organ, tissue, cell, molecule to atom. It is an interesting question to ask, "What kind of information can we obtain when we study at different hierarchical levels?" Before further discussion, I would like to give some definitions first. Physical property is defined as any property that is measurable, whose value describes a state of a system and is also referred to as observable while chemical property is any of a material's properties that become evident during, or after, a chemical reaction; that is, any quality that can be established only by changing a substance's chemical identity (Wikipedia). I think that biological property may be defined as any properties associated with the processes and activities that are unique to life. Generally speaking, we can obtain the physical properties of a system at any hierarchical levels, from atomic level to the level of individual organism. This is because the properties such as force, energy and stability exist for all these hierarchical levels. However, the chemical properties of a system can only be obtained at the molecular level and beyond. The chemical properties of a system are mostly associated with breaking and forming chemical bonds of molecules in chemical reactions. Obviously, research at atomic level cannot obtain any of such properties. Similarly, the biological properties of a system can only be obtained at the cellular level and beyond. This is because, according to the existing cell theory, the cell is the most basic structural and functional unit of life and study at the molecular level and below cannot obtain direct information of life. I think that when we study the living system at the molecular level, we only study the physicochemical attribute instead of the biological attribute of the system because the information of biological attribute does not contain in any hierarchical structures below the cellular level.

Definitely, the physicochemical attribute of the living system is very important. The study of the attribute reveals the characteristics and principles of the living system at the molecular level and has opened the door for us to understand life and laid the groundwork for biological research. For instance, the double helical structure of DNA by Watson and Crick helped us understand the mystery of inheritance. However, we should be aware of the other important attribute of the living system, the biological attribute. The current study in biology usually neglects this basic attribute of the living system or even takes the physicochemical attribute as the only attribute of the system. More importantly, we should also be aware of the limitation of the methodology used in current biological study. For instance, in the research of molecular biology, we inevitably cut the cell apart to study the molecules inside. In such a situation the cell essentially loses its information of life even though its components remain basically intact. Therefore, what we obtain from molecular biology is the physicochemical attribute of the cell but not the biological attribute of the cell. Hence, if a "phenomenon of life" can be explained by physical sciences, then it is essentially the phenomenon of physicochemistry, or the physicochemical attribute of life. Addy Pross put forward the concept of dynamical kinetic equilibrium to describe the living system and claimed that biology can be reduced to chemistry (Pross, 2012). I think that he might just reduce the physicochemical attribute of the living system into chemistry instead of all the attributes of the living system into chemistry. Here I must emphasize that the discussion above does not imply that the research of living system at molecular level is inappropriate or meaningless but only points out that the living system has different characteristics at different levels and combination of all these characteristics at different hierarchical levels certainly helps us understand fully the phenomena of life as a whole.

1.4 The Cell as the Target for the Research of Theoretical Biology

Living organisms in the biological world are diverse in function and structure. Each species has its own characteristics in the hierarchical levels of organization. Darwin may be the first to study systematically for universal theories of biology successfully. The basic unit he studied was the whole individual organism. He investigated the properties of biopopulation composed of individual organisms and obtained some important principles in evolution of species. He did not investigate lower hierarchical levels of organizations than individual organism because of the obvious restriction of the scientific methods and techniques at that time. With development of biological sciences in the 1980s, we have better and deeper understanding of the processes and the structures of living organisms at the cellular and molecular levels. Accumulation of the knowledge makes it possible for us to analyze and explore theoretically the principles and mechanisms of the living organisms at these levels.

What is the appropriate hierarchical level of organization for investigation of universal theories in biology? Structurally, the biosphere may be classified hierarchically as biopopulation, species, individual organism, organ, tissue, cell and molecule. I believe that the cell is the best candidate for theoretical research at the present stage. The reason is simple. At lower hierarchical levels of organization than the cell, there are usually no activities of life. Even though some viruses of nucleic acid and protein complexes are considered to be involved in some kinds of activities of life, they cannot live independently and furthermore no higher hierarchical organizations can be built up from them. Therefore they are not a good candidate for theoretical research. At the higher hierarchical organizations than the cell, the structures and functions of different species diverge tremendously and are difficult to investigate and analyze their universal behaviors and mechanisms. As a starting point for theoretical biology, the first issue we should investigate is the fundamental difference between the living system and the non-living system and their different behaviors and characteristics. Clearly, the cell is the best candidate for these aims because the cell has all the characteristics of the living system and can reflects the fundamental differences between the living system and the non-living system. As the American biologist E. B. Wilson pointed out that all the answers to biology are eventually sought in the cells because all organisms are or once were a cell.

We may ask whether the description of the living system at molecular level is more accurate than that at cellular level. My answer is "Not Necessary". Definitely the description at molecular level is more detailed microscopically. However, is it necessarily true that more detailed means more accurate? My answer is "Not necessarily". It depends on what we are studying. For instance, observation of a cell with a microscope is more accurate than that with naked eyes. However, observation of a football with a microscope is not more accurate than that with naked eyes. Even though the observation of a football with a microscope may give us more microscopic information, the observation with naked eyes gives us better ideas of the characteristics and functions of a football.

As for the question of whether the laws and principles at cellular level can be eventually reduced to those at molecular level, my answer is "NO". According to autonomy of biology, the laws and principles at these two different levels belong to different worlds, one is the living world and the other is the non-living world, and those laws and principles associated with the biological attribute of the living system cannot be reduced any further. Therefore, the answer to the question is "NO". Even though the mechanisms and rules obtained from molecular biology may be helpful in understanding the phenomena in biology, they should not serve as the criteria for judging the validity of the theories at cellular level. The intrinsic properties of the cell are the total effects of the interactions among all its components and they reflect the collective properties of the system as a whole and thus are not anything mysterious such as the vital forces, either.

A complex organism may be regarded as a society of cells. It is usually composed of billions of cells with different types, which display quite different morphologies and perform different functions. The cell is mainly involved in five different biological processes: growth, division, differentiation, senescence and death. These processes are closely associated with life of the cell. Usually, growth increases the materials inside the cell and volume of the cell; division increases the number of the cell and differentiation increases the type of the cell. Senescence and death are two closely related processes for the cell and senescence usually links to death. Theoretical research of biology will explore the principles and mechanisms for these basic processes.

As a living system, the cell is very complex in structure and is involved in vast interactions with the environment. These interactions are strong and play crucial roles in the processes and activities of the cell. However, as the first step in theoretical investigation, we need to study the intrinsic properties of the cell and the fundamental differences between the living system and the non-living system. Therefore, in this book, I mainly study the characteristics and behaviors of the cell as an isolated system. The interactions between the cells and the

On the Living Cell—A Theoretical Exploration of Biology

signals and impacts from the upper hierarchical levels of organizations such as organs and individual organisms will be considered in further studies. Although this is an ideal model in the idealized situation, the study of isolated cells will provide the intrinsic properties of the cell and will demonstrate the difference between the living system and the non-living system and will lay the foundation for further theoretical research of biology.

In addition, in order to concentrate on the intrinsic properties of the cell, unless otherwise specified, the discussion in the book always assumes that the cell is under the ideal conditions, namely: 1) sufficient nutrients; 2) without any interference or damages from the environment.

No organ will be formed... for the purpose of causing pain or for doing an injury to its possessor.

——Charles Darwin

Chapter Two
Survival and Purposefulness in Biology

2.1 Background

The living system and the non-living system are two totally different kinds of systems in the natural world and exhibit entirely distinct behaviors and characteristics. From this chapter to Chapter 5, we will investigate the fundamental differences between these two different kinds of systems. In this chapter, we will study the most essential difference between these two kinds of systems: purposefulness. The term of purposefulness is sensitive in science because traditionally purposefulness and the related terms such as teleology are classified into the category of subjectivity and thus are excluded from the scope of scientific research. Even though purposeful feature is ubiquitously observed in the behaviors and activities of the living system, the term is always avoided in scientific discussion. If it was used, it was only regarded as a convenient way of describing the phenomena of the living system. In this chapter, we will discuss in depth the concept of purposefulness and some related issues such as survival from the viewpoints of philosophy and science.

As both the basic structural unit and functional unit of life, the cell is involved in all biological activities. Are there any activities that have higher priority than the others in its mind? What rules does it follow in its behaviors and activities? In the following sections, we will answer these questions in detail.

2.2 The Axiom of Survival

The Axiom of Survival: To survive in the existing circumstance is the

ultimate goal of the living system in all its processes and activities.

The axiom is proposed to be valid for simple monocellular organisms such as bacteria and yeasts and the cells that form complex multicellular organisms such as human beings. It indicates that the cell takes survival as the most important criterion for its biological processes and activities. In other words, if any processes or activities are detrimental to its survival, they will be terminated or at least modified in order to obtain better chance to survive.

The axiom is supported by ample evidence and observations. For example, bacteria in a solution move towards glucose but move away from toxins because glucose is beneficial to their survival while toxin is harmful to their survival. In order to survive, many bacteria overcome antibacterial treatment by expressing proteins that confer antibiotic resistance, such as efflux pumps or develop the resistance to conventional antibiotics after continuous evolution. They can also change their structure to avoid being killed by antibiotics. Cancer cells undergo metabolic reprogramming to sustain survival and rapid proliferation. Furthermore, cancer stem cells struggle to survive by regulating specific gene expression to resist the external stresses. For example, by enhancing the expression of trans-membrane ATP-binding cassette transporter, cancer stem cells can transfer chemotherapeutic drugs from inside to outside of the cells against the concentration gradient to reduce the drug concentration inside the cells so that they can avoid killing effect of the chemotherapeutic drugs. This is the main reason for multi-drug resistance in cancer (Leonard, et al., 2003)

Some basic biological processes can also be explained by the axiom. For instance, all cells must finish replication of their DNA prior to cell division. This assures that each divided cell receives all the genetic materials necessary to survive. Cells need to control time and frequency of division in order to develop properly for better survival. Obviously, the processes and activities of the cell is always goal-oriented and can be explained by its purpose for survival in the existing conditions, if not directly associated then must indirectly as a means to that end.

Even though the concept of survival has been widely used in evolutionary biology, introduction of survival into theoretical study of functional biology is still necessary because it reveals the fundamental property of the cell and points out the intrinsic difference between the living system and the non-living system.

The axiom is related with two basic concepts: purposefulness and survival. Because purposefulness is a very complicated issue that is related to many aspects in science and philosophy, I will discuss it in section 2.4. Here I will discuss survival first.

2.3 The Relationship of the Axiom of Survival and Cell Death

We do observe cell death. It should be emphasized that the axiom of survival does not mean that cells do not die but only implies that cells do everything possible for them to avoid death. Cell death has three major different forms, apoptosis, autophagic cell death, and necrosis. Also cell death can be classified into passive death and active death. The passive death indicates that the cell is irreparably damaged and thus is killed while the active death indicates that the cell actively participates in its death, essentially a suicide. Most of the suicide occurs by apoptosis (Green, 2011).

Among the various forms of death, apoptosis and active death look contradictory to the axiom of survival and thus need more detailed discussion. According to the axiom, the cell takes survival as its ultimate goal in all processes and activities and thus should not be involved in apoptosis or active death. For further discussion, a more detailed and specific definition of active death and passive death is needed. I think that active death should be the process in which the cell itself triggers the pathway of death and passive death should be the process in which the signals from outside of the cell trigger the pathway of death. According to this definition, even if the cell is involved in the pathway of death, the cell death may not necessarily be the active death. It is the source of the triggering signal that determines whether or not the cell death is active or passive.

Apoptosis usually occurs in multicellular organisms, including animals and plants and maybe in some unicellular organisms such as yeast (Green, 2011). There are some studies on the programmed cell death in the unicellular organisms and ascribed the altruistic suicide to the kin/group benefits but the results and explanations were equivocal and need further investigation (Nedelcu, et al., 2010; Green, 2011). The discussion here is thus focused on studies on multicellular organisms.

A multicellular organism is usually composed of many different kinds of cells and each kind of cells performs its specific functions to keep the integrity of the organism. For instance, our human beings have about 200 different kinds of cells and more than 10^{14} cells in total. All cells in the organism are orchestrated cooperatively and therefore the organism may be regarded as a society of cells. Furthermore, there is a universal phenomenon in biological world that the upper hierarchical organizational structure in an organism has the ability in restraining the behavior of the lower hierarchical organizational structure. For instance, a tissue of the human body has the ability in restraining the behaviors of its component cells. When some lesions occur in a cell, presumably, the cell will do everything possible to repair the lesions and damages for better chance of survival. Sometimes the cell can succeed but sometimes it fails. In the case of failure, the existence of such a cell may not be conducive to the survival of the tissue and even the whole organism. In this case, the tissue or the organism will trigger the protection mechanism to get rid of the cell. Therefore, it is the higher hierarchical organization that forces the cell to die for better chance of survival of the higher hierarchical organization. The axiom of survival has hierarchical meaning for the organisms with different hierarchical structures.

Apoptosis occurs through some pathways inside the cell, which indeed gives an impression that the cell participates actively in death process and is taken as the evidence of suicide. Among various pathways of apoptosis inside the cell, the most frequent pathway is the mitochondrial one in which the outer mitochondrial membrane is disrupted and apoptosis is triggered (Green, 2011). The pro-apoptotic BCL-2 effectors promote apoptosis by causing mitochondrial outer membrane permeabilization (MOMP) while the anti-apoptotic BCL-2 proteins prevent apoptosis by avoiding MOMP. The role of BH3-only proteins is crucial in the process because they can determine the occurrence of apoptosis by regulating these two types of BCL-2 molecules. BH3-only proteins induce apoptosis by activating the pro-apoptotic BCL-2 effectors or by reducing the activity of the anti-apoptotic BCL-2 proteins. Therefore, the BH3-only proteins play a key role in determining the fate of the cell. The BH3-only proteins are expressed and regulated differently in different environment conditions. Therefore, they are the targets of signal transmission in apoptosis and may be regarded as the connector that links the environment to the mitochondrial pathway of apoptosis (Green, 2011). It is obvious that apoptosis in the mitochondrial pathway is triggered by the environmental signals.

The environment of the cell usually deviates from the ideal conditions and the cell is confronted with various physical or chemical stresses such as DNA damage, growth factor deprivation. These stresses usually trigger the mitochondrial pathway of apoptosis. Cells will die through apoptosis when these stresses beyond some threshold values at which cells can no longer maintain their integrity. It is clear that the pathway is activated in response to cellular stress. Apoptosis is the process in which the organism forces the detrimental cells to die for its better chance of survival. In other words, apoptosis is not the voluntary behavior of the cells themselves. Therefore, apoptosis is not contradictory to the axiom of survival in the multicellular organisms.

In addition to cellular stresses, cell death also occurs by developmental cues for both vertebrates and invertebrates. The generation of digits in the vertebrate limbs provides a vivid example. The difference in development of limbs of chick and duck indicates the characteristically different outcomes by apoptosis. The webbed feet of ducks help them swim while chicks cannot swim so webbed feet do not help anything in their life. Such cell death by apoptosis is precisely controlled by the genetic programs designed for the need of the living organisms. In this case, the cells are also forced to die according to the requirement of the genetic programs which are on behalf of the better chance of survival of the whole organisms.

Another common pathway of apoptosis is the death receptor pathway in which extracellular ligands bind the cell surface at some specialized molecules called "death receptors". The death ligands belong to the tumor necrosis factor (TNF) family and the death receptors belong to TNF receptor (TNFR) family

(Green, 2011). When the death receptors are ligated, they will activate caspase-8, which causes MOMP and drives the mitochondrial pathway of apoptosis. The death receptor pathway is an example that the cell is triggered to participate in death pathway by external signals.

Similarly, the inflammasome pathway is induced by signals resulting from infection. In organisms with immune systems, lymphocytes and phagocytic cells can detect an infected cell and instruct it to die in the inflammasome pathway.

Cell death mechanism is also triggered when a cell acquires mutations that make it lose its social inhibition and proliferate as a cancer. The existence of cancerous cells is detrimental to the survival of the whole organism therefore there are some signals that trigger the cancerous cells to die. For instance, tumor-suppressor mechanisms specify the cancerous cells to die, which is similar to the developmental cues (Green, 2011). However, cancerous cells undergo metabolic reprogramming to sustain survival and rapid proliferation. In other words, the difference between the cancerous cells and the other cells is that cancerous cells can build up strong enough resisting power to defend themselves against the death command coming from the higher hierarchical organization.

According to the axiom of survival, I propose a possible mechanism for apoptosis at cellular level. When the cell suffers interventions from the environment such as physical or chemical stresses or DNA damages, the cell is activated in response to such interference. The cell will do its best to resist these stresses and repair the damages occurring inside the cell. However, the stresses may not be resisted entirely or the damages may not be repaired completely, these remanent effects and damages will accumulate and deform the cell. When these effects and damages reach some threshold values, the deformation of the cell will be detected by the guarding system of the organism. The organism will send some signals to trigger the death process or send some death ligands to bind the dead receptors and then apoptosis ensues.

2.4 Purposefulness in Biology

Purposefulness is an important concept in philosophy of science. The research of purposefulness started as early as the time of Aristotle. Aristotle observed the phenomena in biological world carefully and concluded that all these processes were goal directed. In his Final Cause, Aristotle extended this conclusion to the non-biological world and proclaimed that an underlying purpose was associated with all material in both biological and non-biological worlds (Pross, 2012). During the modern scientific revolution in which our understanding and knowledge of the non-biological world improved dramatically, the idea of an underlying purpose in nature was completely dismissed and was replaced by the idea that nature is completely objective. Obviously, such change is mostly based on the scientific knowledge obtained in the non-biological world, more specifically, physical world. Logically, the methodology that has been used now is in principle the same as Aristotle's, namely, extrapolating the conclusion obtained in one field to another without verification. The results of modern scientific revolution have proved that Aristotle's extrapolation was wrong. How about the contemporary one? Let us go back to study in depth some philosophy of science.

Since the scientific revolution physics has achieved great success but the research in biology developed very slowly, therefore the concepts and theories of scientific philosophy were built essentially on the basis of physics and belong to physicalism. Comparatively, biology has little impact on these concepts or theories. Physicalists were essentially reductionists, thinking everything is mechanical and deterministic (Mayr, 2004). They tried to explain all biological processes with the concepts and principles of physical sciences. They also attempted to reduce the complex biological system into the simple physicochemical system and to reduce the concepts and theories of biology into those of physics. For those concepts and processes that cannot be reduced to those of physics, physicalists simply label them as non-scientific or non-existent (Mayr, 1997). Purposefulness is certainly among these concepts. They claimed that because purposefulness does not exist in the non-biological world and cannot be reduced to physical sciences, it should not exist in the biological world either.

With rapid development of biology in recent decades, the limitation of the concepts and theories in physical sciences and traditional philosophy of science

On the Living Cell—A Theoretical Exploration of Biology

became more and more obvious. In order to meet the requirement of the development, we should not restrain the new ideas and new concepts coming from experimental observations and theoretical analysis to fit the framework of traditional theories. Instead, we should modify and extend those outdated theories to agree with the new findings in the biological world. Undoubtedly the purposeful character in biological world exists ubiquitously and undeniably. We should not reject this ubiquitous character just because it is not consistent with the classical philosophy of science. Instead, we should modify and extend the current scientific scope to contain this particular character in the theory of biology. Ernst Mayr analyzed purposefulness and its roles in development of biology with new philosophical perspective, demonstrating that all the goal-directed processes in biological world is programmed processes (Mayr, 2004). It is legitimate to use goal-directedess in the process of evolution and to consider evolutionary processes or evolutionary trend as goal-directed. The goal-directedness in processes are controlled by the programs that may be the products of natural selection in evolution. He used teleonomy to express the goal-directedness controlled by evolutionary programs (Mayr, 2004). Therefore, Mayr demonstrated the purposefulness in evolutionary biology.

Recently, Addy Pross described teleonomy in much more detail but mostly he still confined the discussion within evolutionary biology. Everyday we observe phenomena related to both the living system and the non-living system and we have experienced their great differences. Pross gave a vivid description: "All living things behave as if they have an agenda. Every living thing goes about its business of living—building nests, collecting food, protecting the young, and of course, reproducing. ... We intuitively understand the operation of the biological world, including, of course, all human activities, through life's teleonomic character. ... In the non-living world, by comparison, understanding and prediction are achieved on the basis of quite different principles. No teleonomy there, just the established laws of physics and chemistry." (Pross, 2012) It is obvious that teleonomy is the fundamental difference between the living system and the non-living system. Living systems have an "agenda" and can act on their own behalf. Pross tried to explain teleonomy from the laws and principles of physical sciences. He proposed the important concept of dynamic kinetic stability (DKS) and attributed the intrinsic nature of purposefulness to the energy-gathering capability in the living system, which can free the living system from the thermodynamic constraints for the most stable state in term of energy and can maximize their DKS. To maximize their DKS, the living system needs to consume energy. Such energy can move against the thermodynamic equilibrium and keep the living system in the far-from-equilibrium state (Pross, 2012). He took a car for an example. A replicating entity with an energy-gathering capability is like a car with an engine—it can go uphill instead of only going downhill for a car without an engine.

Definitely, Pross has made a big step forward for our understanding of the basic characteristic of purposefulness from the viewpoint of physical sciences. However, I think this is only part of the story. Purposefulness does not only mean the "uphill" behavior of a car with an engine. It still needs a driver or some directives for a car with an engine to get to the destination. In other words, to accomplish an agenda, the replicating entity with energy-gathering capability is not enough. It still needs decision-making capability to know what to replicate, when to replicate and how to replicate. This is the essential part of accomplishment of an agenda. Where and how can a replicating entity with only energy-gathering capability achieve this information? Obviously, energy-gathering capability is only a necessary requirement for a replicating entity to accomplish an agenda. The critical part of purposefulness is where to go and how to act on its own behalf. In other words, energy-gathering capability is only the necessary condition for fulfillment of purposefulness, and the decision-making capability is the key part of purposefulness. Clearly purposefulness cannot be explained satisfactorily only by the laws and principles of physical sciences. The relationship between biology and physical sciences is the hierarchical levels in nature, with biology having higher level of organization. As pointed out by Niels Bohr in the 1930s, the life phenomena should be explained by the principles that are different from those of physics and chemistry (Bohr, 1933). In fact, purposefulness belongs to the biological attribute of the living system and, as discussed in Chapter 1, thus cannot be explained fully by the principles and laws of physical sciences.

Complexity plays an important part in the characteristics of the living system. As pointed out by Mayr, "There are no inanimate systems in the mesocosms that are even anywhere near as complex as the biological systems of the macromolecules and cells" (Mayr, 2004). Even though the living system and the non-living system are made up of the same "dead" chemical elements, the emergent properties of the living system are tremendously different from those in the non-living system. Purposefulness is one of these emergent properties, reflecting the purposive behaviors and activities of the living system (Mayr, 2004). Although we still do not understand how the combination of the "dead" molecules turns into the living cell that is endowed with purposefulness, its existence in the living world is undeniable. In fact, purposefulness is nothing but the result of holism, namely, the emergent characteristics coming from the interactions among the components, similar to the aquosity of water molecule coming from the combination of hydrogen atom and oxygen atom. Therefore purposefulness is not mysterious at all. It seems mysterious to some people maybe just because it cannot be explained by physical sciences.

To some extent purposefulness has been accepted in explanation of the "ultimate causality" in biology, namely, the field of evolutionary biology. The axiom of survival indicates that purposefulness can also be applied to explain the "proximate causality" in biology, namely, the field of functional biology. Purposefulness is not just a succinct term to describe the behaviors or processes of the living system. It comes directly from our observation and understanding of the living system and is indeed the intrinsic property unique to the living system. Behaving purposefully or not is the fundamental difference between the living system and the non-living system. Therefore, the purposefulness of the living system should be regarded as the most basic and primitive element in the theory of biology and cannot be reduced any further to physical sciences.

Another reason that purposefulness has not been universally accepted in science is the impact of the philosophy of science which is rooted from physical sciences. Traditional philosophy of sciences considers that purposefulness is the subjective thoughts of human beings and should not fall into the category of scientific research because science only investigates objectivity. However, the purposeful character of the living system is so closely associated with its behaviors that we cannot understand the system without the concept of purposefulness. Hence biologists face a dilemma. Just like what Haldane described, "Teleology is like a mistress to a biologist: he cannot live without her but he's unwilling to be seen with her in public."

What should we do and where shall we go? First of all, we should be aware of the constraints of the concepts of traditional sciences on our mind and our ways of thinking. And then we search for possible ways to break through these hurdles and establish a new theory to describe the principles and rules in the biological world. The development of modern sciences has indicated that the subjectivity and the objectivity of the natural world are not completely separated as we might expect previously. Take the measurement in quantum mechanics for an example. In general, the state of a quantum system is a superposition of some eigenstates. A measurement (or observation) of the system causes the wave function of the system to collapse into one eigenstate. In other words, if we want to know at what state a quantum system stays, we have to measure the system. Before the measurement, we only know the probability that the system will stay at each eigenstate, but we do not know which state the system will exactly stay. A measurement includes the physical interaction between the objects to be measured and the measurement apparatus, and psychophysical interactions between the apparatus and the observers (Jammer, 1974). In this case, the objective state depends on the subjective measurement. This situation is vividly illustrated by the well-known paradox of Schrödinger's cat. The paradox says that a cat is put inside a black box in which a specially designed apparatus has 50% chance to kill the cat. When we open the box, the cat is dead or alive with 50% probability for each incident. However, before we open the box, according to the basic principles of quantum mechanics, the cat is at such an uncertain state that it may be dead and may be alive with 50% of probability for each incident, namely, the cat is both dead and alive at the same time (Gribbin, 1984). It does not mean that we do not know the fate of the cat because we do not open the box, but the cat itself is at an uncertain state of life and death due to the fact that we do not open the box. In other words, the objective fate of the cat depends on the subjective action of our observation, which is contradictory to the experience in our daily life. In the quantum world, the subjective observation and the objective states are closely related.

Quantum entanglement is an interesting phenomenon in the microscopic world. No matter how far apart, two quantum particles from a common source engage in the entangled relationship, namely, a particle instantaneously knows the measurement made to the other entangled particle. Experimental results showed that once the quantum entanglement relationship is established, the relationship will be maintained and the particle can distinguish and recognize the other entangled particle without limitation of space and time. This phenomenon cannot be explained by the concepts of classical physical sciences. In fact this characteristic of microscopic particles is similar to the consciousness of human beings. Therefore we may consider the existence of quantum entanglement as an evidence for existence of consciousness for the microscopic particles. We all know the existence of our consciousness. We feel confused just because we cannot describe and explain it by using the concepts of time, space, mass and energy. Further analyses of quantum measurements showed that consciousness cannot be reduced any further. Now, more and more people believe that consciousness is another basic and independent characteristic of the substance in the natural world, just like time, space, mass, and energy. Just like we cannot explain time with space, we do not expect that consciousness can be explained by any of the basic physical quantities such as time, space, mass or energy.

Let me take schizophrenia for another example. Schizophrenia is a mental disease characterized by abnormal social behaviors and is ascribed to deficits in cognitive abilities. Previously, the diagnosis is mostly based on observed behaviors and reported experiences and no objective tests are usually available. Studies with brain imaging technologies such as fMRI and PET showed that the disease is associated with abnormalities in brain structures. This example clearly shows that the subjective human behavior is closely associated with the objective brain structures. Similarly, I expect that the "subjective" purposefulness will find its objectivity in the future research of brain structure.

From all the above examples, we can see that with the progress of science, especially biological sciences, the boundary between the traditional definition of subjectivity and objectivity becomes increasingly blurred. We may expect that subjectivity and objectivity will be more and more closely linked and mutually influenced and cannot be separated completely. Perhaps subjectivity and objectivity could be unified somehow in the dual attributes of the living system, namely, the traditional objectivity may be displayed in the physicochemical attribute while the traditional subjectivity may be displayed in the biological attribute. I think that if we want to investigate the essence of life and search for the fundamental rules unique to the biological world, we inevitably include purposefulness in our theory of biology. Without purposefulness, we may only study the physicochemical attribute of the living system.

With introduction of purposefulness into the theory of biology, we may predict some behaviors and activities of the living systems. For instance, we may predict the trajectory of bacteria towards glucose in the solution and may explain the salmon's spawning migration.

It is quite understandable for the attempts to explain the characteristics of purposefulness and the axiom of survival with the principles and concepts of physical sciences. Methodologically speaking, however, such endeavors are most probably doomed to failure and disappointing, just like the attempt of understanding the principle of constancy of the speed of light in the special theory of relativity with the concepts and principles of the Newtonian physics. Even though the principle of constancy of the speed of light is contradictory to that in Newtonian physics, they still belong to the inanimate world. The difference between the principles in the animate world and the inanimate world are expected to be much more tremendous, therefore it is not surprising if the principles in the animate world cannot be explained by those in the inanimate world. For instance, the hard problem of consciousness, namely the experience and feeling, cannot be explained by physical sciences because it is beyond the scope of physical sciences. Rosenberg believed that the teleological explanation may not be completely reduced to non-teleological interpretation in practice. Generally, the biological attributes of the living system should not be expected to be explained by the principles of physical sciences.

2.5 The Significance of the Axiom of Survival

Generally speaking, survival has two implications, one is the survival of species which is closely associated with the transfer of genetic information from one generation to the next and belongs to the category of evolutionary biology; the other is the survival of individuals which is related to the behavior of the individuals and belongs to the category of functional biology. The survival of individuals is the basis for the survival of the species and the survival of species guarantees the survival of the individuals. The two are mutually beneficial and closely related. The survival of the species is believed to follow the natural selection in which the fittest survives. The axiom of survival points out that the behaviors and activities of the living system are based on the ultimate purpose for survival in the existing circumstances.

Goal-directedness of behaviors and activities ubiquitously exist in the biological world and has been studied from the time of Aristotle. Teleology was used to describe the goal-directednss in cosmic world, including non-biological and biological worlds and teleonomy was used to explain the goal-directedness in evolutionary biology. All these studies are mostly confined in philosophical discussion. Here purposefulness is used to explain the goal-directedness in functional biology and to describe the concrete and specific behaviors and activities of the living system.

What is the significance of the introduction of purposefulness into functional biology by the axiom of survival? First of all, introduction of purposefulness reveals the fundamental difference between the living system and the non-living system. The purposefulness here is not merely the convenient rhetoric in describing the observed phenomena; instead it is indeed the intrinsic properties that the living system is endowed with. It is a concrete manifestation of the autonomy of biology. Newton's First Law of Motion introduces inertia into physics, which is the intrinsic property of bodies in all processes of motion. Similarly, the axiom of survival introduces purposefulness into functional biology, which is the intrinsic property of the living systems in all biological processes and activities. Secondly, introduction of purposefulness indicates that the living system follows some other rules that are totally different from those for the non-living system. We are required to establish a brand new framework of theories for the living system. Thirdly, searching for survival is a fundamental principle in the biological world. Without instinct of survival, the living system will have no rationales to follow in its behaviors and activities. Therefore, the axiom of survival is the basis of biological theories. Without the axiom of survival, there will be not any biological theories possible. The above implication of the axiom is profoundly and easily understood if we apply the axiom to the special biological world, the human societies, in which all human laws and regulations would lose their basis and significance if the human beings did not take survival as their ultimate purpose.

2.6 A Definition of Life

Biology is a discipline of life phenomena and aims at revealing the basic principles of life. Therefore, "what is life" is a fundamental question in biology. This question has been attracting human beings to answer since the ancient time. Until now we still have not obtained a clear and satisfactory answer to it. In the past 200 years, there have been a lot of definitions of life, which are mainly based on biological functions relevant to life such as genetics, reproduction and metabolism. Marcello Barbieri listed many definitions of life in his book (Barbieri, 2003) and I just give some of them in the following:

"I suggest that these three properties—mutability, self-duplication and heterocatalysis—comprise a necessary and sufficient definition of living matter." (Norman Horowitz)

"Living beings are teleonomic machines, self-constructing machines and self-reproducing machines. There are, in other words, three fundamental characteristics common to all living beings: teleonomy, autonomous morphogenesis and invariant reproduction." (Mond & Jacques)

"A living system is an open system that is self-replicating, self-regulating and feeds on energy from the environment." (Sattler, R.)

"Life is a chemical system capable to replicate itself by autocatalysis and to make errors which gradually increase the efficiency of autocatalysis." (Brack & Andre)

There are also some definitions that are based on structure. For example, "Life is an emergent phenomenon arising as the molecular diversity of a prebiotic chemical system increases beyond a threshold of complexity." (Kauffman, 1995) Stuart Kauffman further explained that "Life, in this view, emerged whole and has always remained whole. Life, in this view, is not to be located in its parts, but in the collective emergent properties of the whole they create. ... The collective system is alive. Its parts are just chemicals." (Kauffman, 1995)

These definitions may not give a satisfactory answer to life yet. For instance, there is no difference in structural complexity between a live fish and a just dead fish but they display different characteristics in their behaviors and belong to different kinds of systems.

In order to figure out what life is, I think that we need to know the essential difference between the living system and the non-living system. From the discussions in the previous sections, we know already that the difference is purposefulness. Here I propose a definition of life:

Life is the ability to behave purposefully.

The above definition may not completely cover all the aspects of life but, I think, it grasps the essential point of life. Furthermore, it is quite straightforward and easy to identify whether a system belongs to living system by direct observation. Take virus for an example. A virus cannot replicate and metabolize by itself and needs a host cell to survive. But experimental results indicated that a virus uses a lot of tricks to invade the cell and survive there. Obviously, it behaves purposefully. According to the above definition, a virus is classified into the living system. However, in accordance with commonly used definition of life, "a living system should have the ability to self-replicate and metabolize", it is difficult to decide whether it should be classified into the living system.

Life is order, death is disorder.

———Bo C. Malmstrom

Senescence has no function——it is the subversion of function.

———Alex Comfort

Chapter Three
Orderliness and Senescence of the Cell

3.1 Background

We all have observed senescence in living organisms and believe that most, if not all, of the living organisms end up with senescence. The observations of senescence in daily life and research experiments gave us an impression that senescence is a spontaneous process by the living organisms themselves. This situation reminds me of the similar observations of physical systems in the inanimate world: An object without any driving forces decreases its speed and will finally come to a stop. Such observation led Aristotle to reach the conclusion that a force must be exerted on an object in order to keep it moving at uniform velocity. This conclusion had been accepted for about two thousand years until Galileo, based on his experimental results, pointed out that it was the friction not any intrinsic properties of the object that made the object slow down and stop. This idea was generalized by Newton and stated in the his First Law of Motion: "Every body continues in its state of rest, or of uniform motion in a right line, unless it is compelled to change that state by forces impressed upon it." In other words, every body has its inertia to resist to any attempts which try to change its state of motion. How about the senescence of the living system?

In this chapter, I will study the nature of orderliness of the living system and the characteristics of senescence and finally the relationship between orderliness and senescence of the cell.

3.2 Orderliness of the Living System

The living system is endowed with high complexity and exquisite orderliness. It is capable of self-assembling in structure, self-regulating in physiological activity, and self-replicating in proliferation. These characteristics of the living system are attributed to its complexity and orderliness. In addition, the living system is an open system and it can keep good homeostasis despite continuous exchange of large amount of matter and energy with the environment. The reason for this feature is the same, high complexity and exquisite orderliness of the system. Another characteristic of the living system is its hierarchical structure. It is all these unique features of the living system that make it autonomous and divergent from the non-living system. For example, the human body is a living system composed of cells, tissues and organs. He can grow, develop, proliferate and metabolize. The hierarchical structures of the human body are highly orderly and the physical and chemical processes and physiological activities inside the body are highly cooperative and well orchestrated. The orderliness of both the organization and the structure increases when human body grows. Therefore the whole human body is an orderly system far away from the thermal equilibrium. If the orderliness of the human body is destroyed and its entropy increases, then life cannot be maintained.

The orderliness of the living system is generally recognized and accepted. However, there are different explanations for the nature and source of the orderliness. If the living system is regarded as a thermodynamic system, so the orderliness of the system needs energy and "negative entropy" to maintain; if the living system is taken as a mechanical system like a pendulum clock equipment, then the thermal motion of molecules is not strong enough to destroy or change the conformation or structure of the system (Schrödinger, 1944), therefore the orderliness of the living system does not need energy or "negative entropy" to maintain. As for the nature and source of orderliness, Schrödinger proposed two different mechanisms: 1) "order from disorder" produced from statistical mechanism and, 2) "order from order" produced by a new type of physical law (or non-physical law or super-physical law) (Schrödinger, 1944).

Chapter Three Orderliness and Senescence of the Cell

Considering the fact that the living system is an open system with orderly structures, it is usually regarded as a "dissipative structure". Dissipative structure was proposed by Ilya Prigogine to explain the observed orderly structures that are induced by the statistical fluctuation when the thermodynamic systems are far away from the thermal equilibrium and some of the parameters reach the threshold values (Prigogine & Nicolis, 1971). The orderliness is sometimes called "order through fluctuation". Even though there are some similarities between the living system and the dissipative structure in thermodynamic systems, such as an open system with exchange of energy and matter with the environment and possessing orderly structure, there are still some fundamental differences between the two. The dissipative structure occurs only when the thermodynamic systems are far from the thermal equilibrium and some of the parameters are near the threshold values so that the thermodynamic fluctuation may be amplified. However, the living system such as a living cell possesses the orderliness once it comes into being, which does not need fluctuation or any other factors to induce. The orderliness increases as the individual grows. Furthermore, the orderliness of the living system enables the living systems such as a bacterium effectively synthesize and distribute different kinds of proteins and other molecules inside the cell, and also a fertilized egg in mammals correctly develops into an organism, all of which are presumably instructed by the information from the genome inside the cell. Therefore, I think that the orderliness of the living system, which usually has the hierarchical structure and is derived from the long process of natural selection, should be essentially different from the dissipative structure induced by statistical fluctuation. The orderliness of the living system should be an intrinsic characteristic of the system by its own.

Schrödinger explained the concept of "order from order" by considering the orderliness of the living system as that of the pendulum clock equipment. He further illustrated the relationship between the clockwork and organism: "Now, I think, few words more are needed to disclose the point of resemblance between a clockwork and an organism. It is simply and solely that the latter also hinges upon a solid—the aperiodic crystal forming the hereditary substance, largely withdrawn from the disorder of heat motion." (Schrödinger, 1944) Stuart Kauffman introduced the concept of "order for free" and proposed that "much of the order in organisms may not be the result of selection at all, but of the spontaneous order of self-organized system. Order, vast and generative, not fought for against the entropic tides but freely available, undergird all subsequent biological evolution. The order of organisms is natural, not merely the unexpected triumph of natural selection." (Kauffman, 1995) He further discussed the essential features of the cell such as the homeostatic stability of the cell (e. g. the biological inertia that keeps a liver cell from turning into a muscle cell) and suggested that the orderly structures in the living system come from the process of self-organization such as in the process of ontogeny (Kauffman, 1995).

I agree with the general idea of "order from order" for the living system because it moved one step away from the traditional concept of thermodynamic equilibrium. But I think that the living system and the mechanical system are two different systems, with the former being dynamic and the latter static. Accordingly, the orderliness of the two systems should be also different. First, the source of orderliness of the two systems is different. The orderliness of the mechanical equipment comes from the designs of craft masters or watchmakers while the orderliness of the living system may be obtained by the process of self-organization proposed by Kauffman. Second, the mechanism for maintaining the orderliness of the two systems should be different. The orderliness of the mechanical equipment is maintained because the thermal motion is not strong enough to destroy the structures of the equipment. However, the living system is a dynamic system with many interacting components inside. The orderliness of the system may be maintained by the mechanisms achieved in the process of evolution.

How is the orderliness of the living system maintained? The living system is an open system and exchanges matter and energy continuously with the environment. Based on the principles of thermodynamics, the orderliness of the living system is maintained by energy and "negative entropy" obtained by the living system from the environment. However, as far as I know, until present

Chapter Three Orderliness and Senescence of the Cell

there is not yet direct experimental evidence to show that the energy or "negative entropy" is used for maintenance of the orderliness of the living system. Based on the following discussion, I propose that the orderliness of the living system comes from the intrinsic property of the system itself and does not need anything from outside to maintain, namely, "keeping order for free". On the contrary, interventions from outer environment are needed to change the orderliness of the living system.

Indeed, the living system needs nutrients to maintain its biological activities. But is the energy or "negative entropy" obtained from the nutrients or food for maintaining the orderliness of the living system? When the food is made up of biological macromolecules and the wastes are inorganic small molecules, the human body surely receives some "negative entropy" because in general, the biological macromolecules are apparently more orderly than the inorganic small molecules. However, from the viewpoint of maintenance of human life, the oxygen and water molecules are more important than food. In this case the human body does not get any "negative entropy" because the oxygen molecules and water molecules are not more orderly than the molecules of the wastes.

Hibernation of animals may provide a convincing example for demonstrating the effect of exchange of matter and energy between outside environment and the living system and may help us understand better the maintenance of the orderliness of the living system. Some hibernating animals can survive in more than 100 days without eating or drinking during hibernation season. In other words, these animals almost do not obtain energy or matter from the environment but can still keep their life (presumably, keep the orderliness of their organizations and structures) by means of extremely low metabolism. This phenomenon shows that when the intake of external matter and energy is reduced, the living system may adjust the level of metabolism to keep the orderliness of the organizations essentially intact. Therefore the orderliness of the living system does not necessarily need the external energy or "negative entropy" to maintain and energy is most probably used only for the biochemical reactions and activities. This situation may be analogous to the operation in a factory. The machines in a factory have already been set up (certainly the setup of the machines needs energy) and energy and materials are needed to operate these machines and to make products. When energy and materials are reduced, the machine will work slowly but the setup of the machines in the factory remains the same. The above discussion may lead us to reach the conclusion that the intake of matter and energy is used for metabolism inside the living system but not for the maintenance of the orderliness; therefore orderliness is intrinsic to the living system.

Schrödinger applied similar argument to the pendulum clock at zero absolute temperature (Schrödinger, 1944). The thermal motion of the air molecules around the pendulum clock impedes the periodic movement of the pendulum and produces heat and the clock needs energy to keep running. However, such thermal motion is not strong enough to destroy the orderly structure of the equipment of the clock. In other words, the thermal motion of the air molecules cannot change the orderliness of the equipment of the clock. The energy is needed only for the running of the clock. As argued by Schrödinger, the orderliness of the pendulum will change only when the equipment was heated to melt.

The laws of thermodynamics are only applicable to thermodynamic systems. The thermodynamic systems are featured with random thermal motion of a large number of molecules. Is the living system a thermodynamic system? Take the cell for an example. Water molecules and inorganic small molecules such as cations occupy most portions of the thermal motion inside the cytosol. However, most of biological macromolecules construct various orderly structures inside the cell such as the organelles therefore these macromolecules are not involved in the random thermal motion. In addition, the orderly structures within the cell are quite stable under physiological conditions and therefore will not be damaged by the thermal motion of the molecules inside the cell. There are also some orderly movements inside the cell such as cyclosis in which organelles, some nutrients and metabolic wastes may be transferred to the appropriate locations in cytoplasm. Under normal physiological conditions, proteins such as transcription factors and enzymes inside the cell spontaneously adopt the orderly conformation and the thermal motions inside the cell cannot change such orderly conformations into the disordered conformations such as the random coiled conformations. There are indeed some free amino acids and nucleotides inside the cell but many of these biological monomers are involved in the high-speed organic synthesis and should not participate much in the thermal motion. Therefore, the cell is different from the usual thermodynamic systems. As stated by Schrödinger, "The living organism seems to be a macroscopic system which in part of its behavior approaches to that purely mechanical (as contrasted with thermodynamic) conduct to which all systems tend, as the temperature approaches the absolute zero and the molecular disorder is removed" (Schrödinger, 1944). If the cell has to be regarded as a thermodynamic system, the orderly structures such as organelles inside the cell should serve as the boundary of the thermodynamic system, which are not involved in thermal motion. The above discussion demonstrated that even though the thermal motion of the molecules inside the cell may increase the entropy of the system, it does not change the main features of the basic orderly structures inside the cell.

It should be noted that the living system on the earth is an open system usually at constant temperature and constant pressure. According to the theory of thermodynamics, the most stable state for such situation is the one with minimum of Gibbs function but not with maximum of entropy. More specifically, Gibbs function G is defined as

$$G = U + pV - TS \qquad (3-1)$$

Where U, p, V, T and S are the internal energy, pressure, volume, absolute temperature and entropy of the system, respectively.

Usually, when S increases, U also increases. The net effect for decreasing G depends on T. When T is big, increase of entropy S may decrease G. However, when T is small, increase of entropy S may increase G. In other words, at constant temperature and constant pressure, the most stable state for the living system is not necessary the one with maximum of entropy.

The living system is featured with orderly biochemical processes at high speed such as synthesis of protein and DNA molecules. Obviously, these orderly syntheses increase orderliness of the system and thus decrease the entropy of the system. On the other hand, degradation of damaged proteins and unwanted proteins increases the amount of free amino acids and thus increases the entropy of the system. In comparison with the usual thermodynamic system, the living system is involved with disordered motion and orderly synthesis, the net result to the orderliness of the system depends on the contribution of these two different kinds of processes.

3.3 Hypothesis of Orderliness of the Cell

The cell is a complex but exquisite and orderly system. Although the cell is a dynamic system and its orderliness is difficult to define quantitatively, the existence of this characteristic is undeniable. The orderliness of the cell may be defined qualitatively as the orderly arrangement of organizations inside the cell and the orderly networks of their interactions. For example, the cytoplasm is not homogeneous and contains many highly orderly structures such as organelles like mitochondria, ribosomes, lysosomes. Cytoskeleton, composed of microfilaments, microtubules and intermediate filaments in eukaryotic cells, forms a complex network for interlinking filaments and tubules, working for accurate positioning and migration of organelles. As a whole with many orderly structured components, the cell orchestrates the processes of growth, proliferation, differentiation and other physiological activities. During these processes, there are many well-coordinated biochemical reactions inside the cell, which occur in the correct timing in response to the requirement of biological functions. Therefore, orderliness plays a vital role in biological processes of the cell. The following hypothesis illustrates the intrinsic property of the cell in orderliness.

Hypothesis of Orderliness: The cell has the tendency to maximize the orderliness of the system.

The hypothesis points out that the processes of the cell have the direction heading the orderliness of the cell to increase. When the cell reaches the state with its maximum of orderliness at given conditions, the cell will keep in this state unless there is some interference from the environment forcing it to change. Due to the stresses or other interference from the environment, the cell may decrease its orderliness. The hypothesis indicates that when the stresses or the interference is withdrawn or reduced, the cell has the tendency to increase the orderliness of the system and try to restore to the original state of the orderliness. It should be emphasized that the hypothesis does not mean that the cell can always reach the maximal orderliness possible for the system. The hypothesis only implicates that the cell can reach the maximal orderliness allowed for the cell to obtain in the existing conditions. Therefore the maximal orderliness that the cell can obtain may vary with the environment of the cell.

This hypothesis shows another fundamental difference between the living system and the non-living system. To a non-living system, it has the tendency to reach the most stable state such as the state with minimum energy (*e. g.* for dynamic systems) or maximum entropy (*e. g.* for thermodynamic systems). Comparatively, to a living system it has the tendency to reach the best function state with the maximum orderliness (Here it is assumed that the higher the orderliness, the better function.). In other words, the state with the maximum orderliness is the most favorable state for the living system, which does not need input of energy or "negative entropy" to maintain. The hypothesis is supported to some extent by Schrödinger, "Life seems to be orderly and lawful behavior of matter, not based exclusively on its tendency to go over from order to disorder, but based partly on existing order that is kept up" (Schrödinger, 1944).

The hypothesis has two implications. The first is that the processes in the cell have a direction which points to the increase of the orderliness of the system spontaneously. Considering from the philosophical viewpoint, the hypothesis is analogous to the zero-force evolutionary law in which the evolutionary system has the tendency to increase its diversity and complexity (McShea & Brandon, 2010). The second is that there is a driving force coming from the internal of the cell to resist any interference from the environment. In this respect, the hypothesis is analogous to the Newton's First Law of Motion in which every body has the capability to keep its state of motion, which is described by the inertia of the body. Similarly, every cell has its capability to resist the interference from the outer environment by means of some corresponding biochemical reactions. The hypothesis indicates that the living system is endowed with similar capability as the non-living system to resist to any interference that attempts to change the states of the system.

The above discussion leads to the following corollary of the hypothesis of orderliness:

Corollary 1: The cell has the capability of resisting the interference from the environment.

The interference from the environment such as physical or chemical stresses may decrease the orderliness of the cell. When the orderliness of the cell decreases, presumably the function of the cell decreases accordingly. In this case, the corollary implies that the cell has the tendency to resist this interference and to attempt to correct the deviation in orderliness and restore to the original orderliness and functionality.

The intrinsic property of the cell of resisting interference may be described by a new concept termed "viability". The viability of the living system may be defined as its capability of resisting the interference from the environment. The stronger the capability of resisting the interference, the greater the viability of the cell. Obviously, the viability should be determined by its genotype and phenotype and should be reflected in its behaviors in biological processes and its lifespan. For instance, different kinds of cells in the same organism may have different viability and the cells of different species can also have different viability.

According to the present cell theory, all cells come from the existing cells by cell division. Therefore, the orderliness of the cell may be considered as the intrinsic property resulting from biological evolution. Essentially, all cells have their own degree of orderliness. For a given cell, the degree of orderliness is closely associated with its function. It seems reasonable to assume that the higher degree of orderliness of the cell, the better function the cell can perform. In other words, when the orderliness of the cell reaches its maximum, the cell can perform function the best. It may be predicted that the cell should have a transition period after cell division. In order to improve its functionality, the freshly divided cells need to adjust its internal structure to increase its orderliness and change from its nascent state to the mature state which has higher orderliness.

3.4 Senescence of the Cell

For more than 150 years there have been heated arguments about senescence. In order to have effective treatment for age-related diseases and to extend healthy lifespan, we need to investigate the origin and mechanisms of senescence. Even though there have been a lot of theories about the origin and mechanisms of senescence, the conclusions of these theories are still quite controversial.

Senescence is a universal phenomenon in the biological world and occurs after some period of time in the life of organisms, with decrease in their functions in physiological activities and their ability to response to stresses. It is usually referred to the irreversible process in which the organisms degenerate their structures and components along with increase in age. Cellular senescence is a process in which the cells stop dividing and are accompanied with structural abnormalities and physiological dysfunction. The senescent cells mainly display reduction in their ability both to adapt to environmental changes and to maintain homeostasis inside. Structural changes usually include decrease in cell volume, reduction in intracellular water molecules, hardening of protoplast, cell shrinkage and loss of normal morphology. In the meantime of the protoplasm change, pyknosis occurs in the nucleus and its structure becomes unclear; the ratio of nucleus to cytoplasm reduces or even the nucleus disappears (Campisi, et al., 2013).

There are many theories about the mechanism of cellular senescence, which can be classified into two different schools: error theory school and genetic/programmed theory school. The error theory school indicates that senescence is caused by a variety of errors in the cell, including the macromolecule cross-linking theory, the free radical theory, the somatic mutation and DNA repair theory and also the waste product accumulation theory. The macromolecule cross-linking theory considers that excess macromolecular cross-linking is the main cause of senescence. DNA cross-linking and collagen cross-linking can damage the function of the cell. The oxydation of unsaturated fatty acids induces the cross-linking between lipoproteins and thus decreases fluidity of the membrane. The free radical theory demonstrates that the free radicals with strong chemical reactivity can attack the biological macromolecules such as DNA, proteins and lipids in the organisms, which causes damages such as DNA breakage, cross-linking and hydroxylation of bases, and also protein denaturation. The somatic mutation and DNA repair theory elucidates that accumulation of induced and spontaneous mutations causes the loss of functional genes and thus decreases the synthesis of functional proteins, which leads to the senescence and death of the cell. For instance, the senescence symptoms of young mammals induced by radiation resemble those of normal aged individuals.

The genetic/programmed theory school believes that senescence is a natural evolutionary process determined by genetics, including the replicative senescence theory, the programmed senescence theory and longevity genes theory. Among the above theories, the replicative senescence theory is the most popular in illustrating the mechanism of senescence. The theory is based on the shortening of telomeres during DNA replication. The mechanism of telomere shorting is simple. The DNA polymerase needs an RNA primer to start its synthesis and the positions left by the primer cannot be filled by DNA thus the telomere becomes shorter after each replication of DNA. Even though it seems straightforward that telomere shortening play an important role in *in vitro* cellular senescence, the effect of telomere shortening on senescence of *in vivo* organisms may not be as simple as we thought because the activity and mechanism of telomerase *in vivo* are quite different from those *in vitro*. For instance, unicellular eukaryotes such as ciliates (Greider & Blackburn, 1989; Collins, et al., 1995; Lingner & Cech, 1996) and yeasts (Counter, et al., 1997; Nakamura, et al., 1997) have high telomerase expression *in vivo*, which can compensate for the telomere shorting in DNA replication. Rainbow trout and the lobster were found to have high telomerase activity in all organs examined, even during adult life (Klapper, et al., 1998a, b).

Hayflick was the first to describe the finite replicative capacity of fetal fibroblasts in culture and to interpret these findings as demonstrating cellular senescence *in vitro*. It was further claimed that cellular senescence *in vivo* is relevant to cessation of cellular growth in culture. The interpretation led to the conclusion that the number of cell division is the limiting factor that serves as a clock for determining the lifespan of a multi-cellular organism. When the telomere length is shortened to reach some threshold values, the cell cannot divide any more.

Despite many studies on replicative senescence, the relevance of *in vitro* studies to *in vivo* senescence has been controversial (Cristofalo, et al., 2004). For instance, one of the major supports for direct relationship of replicative senescence with cell senescence *in situ* was the putative decline in the replicative lifespan of skin fibroblasts (and other cell types) in culture as a function of donor age. However, studies using healthy donors of different ages demonstrated that there was large variability in association of donor age with replicative lifespan for human fibroblasts in culture and did not show statistical significance (Cristofalo, et al., 1998). Another confounding issue relates to the "telomere hypothesis of aging", which proposes that replicative aging may be regulated by telomere shortening. Experimental results showed that telomere length was highly variable in multiple clones established from a single individual (Allsopp and Harley, 1995). Therefore, the above data did not support the direct relationship between donor age and average telomere length *in vitro*.

Gershon and Gershon listed many observations to illustrate no evidence for a "Hayflick limit" in proliferative capacity of cells *in vivo* but substantial evidence that points to lack of proliferative limits *in vivo* instead, demonstrating that what was called "proliferative senescence" could not and should not be used as a model for *in vivo* senescence (Gershon & Gershon, 2001). Harrison conducted a series of experiments of transplanted bone marrow cells and observed that proliferative capacity exceeded that required for the lifespan of a mouse and moreover, in most experiments cells of old individuals proliferated as well as young individuals when transplanted into young recipients (Harrison, 1985). Gershon and Gershon proposed that the "Hayflick phenomenon" was a result of abuse of *in vitro* cells and suggested that the telomere theory of aging be considered irrelevant. Without such abuse the *in vivo* systems have the proliferative capacity that far exceeds the need for maintenance of cellular proliferation throughout the lifespan of the organism (Gershon & Gershon,

2001).

Interestingly, some immortalized cell lines have no detectable telomerase activity but still have long telomeres. This observation may imply that there is an alternative mechanism to lengthen telomeres (Biessmann and Mason, 2003; Bryan and Reddel, 1997; Dunham, et al. 2000). Therefore the mechanism of telomere shortening is complicated *in vivo* and the Hayflick limit may not be a universal standard for testifying senescence.

In summary, the above discussion indicated that telomere shortening is not the only counting mechanism in replicative senescence. Depending on the experimental models, there are other mechanisms that play a role in limiting the proliferative potential of normal cells (Cristofalo, et al., 2004).

Two models of senescence have been proposed from the viewpoint of evolutionary biology (Goldsmith, 2014). The Medawar's model indicates that the genetic drift and mutation accumulation lead to the loss of late-acting beneficial genes or the appearance of late-acting harmful genes. The Williams' model considers that senescence comes from the pleiotropic effect of some genes that are beneficial early in life and then harmful at later ages. Some major questions are still open concerning the precise genetic mechanisms and specific genes underlying the evolution of senescence.

There are two fundamentally different theories about the origin of the process of senescence. In the programmed aging theory, senescence is genetically programmed intentionally because living too long produces evolutionary disadvantages and therefore after some periods of time, aging programs are turned on to accelerate the buildup of damage and decrease the capacity for repair. In the non-programmed aging theory, aging is caused by the accumulation of damages such as those by reactive oxygen species. Once the accumulation exceeds the threshold value, the process of aging will start. There are supporting experimental results for both theories (Goldsmith, 2014; 2015) and the question of the origin of senescence remains controversial.

There are a lot of confusions in the experimental results of the study of senescence. These confusions mostly result from the great differences in the conditions used in the study such as the *in vivo* and *in vitro* conditions. For instance, although the biological characteristics under arteriosclerosis, prostate hyperplasia and other pathological conditions are similar to those of senescence of *in vitro* cells, the majority of markers found in *in vitro* cultivated cells have not been validated in the study of senescence *in vivo*. In addition, the telomeres in the chromosome of *in vitro* cultivated somatic cells of human beings shorten after each cell division and the number that the cell can divide is confined to the Hayflick limit. However, the telomeres in the chromosomes of some species of mice keep a considerable length in their life and no obvious shortening has been observed.

Why are the *in vitro* cells and the *in vivo* cells enormously different? Research shows that the main reason for the difference is the very different environment for the *in vivo* cells and the *in vitro* cells. For example, under general cultivation conditions, cells are incubated in 20% of oxygen concentration, much higher than the physiological condition (3%); in addition, *in vitro* cultured generation transfer of cell needs trypsin-treated cells. Comparatively, *in vivo* cells may not encounter this situation. Trypsin treatment not only disrupts the intercellular junctions but also digests the receptors on the cellular surface. Furthermore, the biggest difference may lie in the fact that the *in vivo* cells are in the three-dimensional physiological environment, while the cultured cells used for senescence study are often only in two dimensions. This makes big difference for the cells in receiving various signals and undoubtedly affects the behaviors of the cell. Until now, no experimental evidence suggests that *in vitro* replicative senescence or premature senescence induced by stress is directly related to senescence in tissues or individuals *in vivo*. It is indeed a great challenge to investigate the mechanism of senescence *in vivo* with the *in vitro* cells.

3.5 Orderliness and Senescence

According to the generally accepted definition, the process of cellular senescence begins when the functions of the cell decrease below some threshold values. Most of the cells change their functions when they senesce (Campisi, 2000). Structure and function are two important aspects of the living system and closely related. When the structure of the cell is damaged, the function of the cell will be disrupted accordingly. Obviously, the orderliness of the cell decreases at the same time. Therefore, the decrease in the cellular function is due to the decrease in cellular orderliness.

The cell is an open system and is involved in interactions with the environment and experiences interference from the environment all the time. The interference such as physical and chemical stresses may disrupt the orderliness of the cell and decrease the orderliness of the system. When the effect of this disruption accumulates to some extent, the cell will deviate from its normal pathway in the biological processes and its functions decline accordingly. Therefore, cellular senescence is triggered to begin. However, every cell has its capability to resist to the interference from the environment and the capability varies among different cells. The difference in such capability results in different onset of senescence for different cells.

We observed in our daily life that any moving non-living objects will slow down and finally stop if there is not force acting on them. However, Newton's first law of motion points out that the objects have the inertia to keep their state of motion. It is the friction that changes their state of motion and makes them slow down and stop. Similarly, in the living world, any organisms will experience senescence and death. What is the causality for these ubiquitous processes in the living world? Are there any driving forces that keep organisms away from senescence and death?

Normal functions of the living system depend on the appropriate orderliness of the system. On the other hand, senescence of the living system is characterized by degeneration in both structures and functions. It is well accepted that senescence begins when the damages inside the cell accumulate beyond some threshold values (Seregiev, et al., 2015). This phenomenon may be explained by the hypothesis of orderliness. As discussed in section 3.3, the orderliness of the cell includes the orderly arrangement of the organizations in the system and the orderly networks connecting all the organizations. The interference and stresses from the environment may cause damages in the system and disrupt the orderliness of the system hence intervene the biological processes. According to the hypothesis of orderliness, the cell has the intrinsic property to increase its orderliness. If these damages are not fatal instantly, the system has the capability of resisting any further damages occurring in the system. In this case, the system can still keep in the normal biological pathways and perform the appropriate physiological functions. However, when accumulation of the damages exceeds the threshold values, the disruption will exceed the tolerance of the system and the orderliness of the system will collapse. Consequently, the physiological processes of the system will deviate from the normal pathways and senescence ensues. The situation is analogous to moving an object at rest in the non-living world. The object will not change its state of motion if the force exerting on the object does not overcome the static friction. However, once the driving force exceeds the maximum static friction, the object at rest will move forward and speed up along the direction of the driving force.

The above discussion not only explains the mechanism of cellular senescence but also leads to the following corollary of the hypothesis.

Corollary 2: Cellular senescence begins when the orderliness of the cell decreases below some threshold values.

From this corollary, it is obvious that the hypothesis of orderliness supports the error theories of senescence because the cell has the intrinsic property to maintain its order for free. Therefore, senescence of the cell does not come from the genetic programs inside the cell but can only result from the interference and damages from the environment.

The cell may suffer many kinds of external stresses, which lead to the abnormalities in the cell such as the presence of excessive free radicals and accumulation of metabolic wastes. The consequence of the abnormalities includes DNA breakage, protein denaturation and loss of base pairing specificity of nucleotides. All these abnormalities reflect the destruction of orderliness of the cell. For instance, denaturation of protein can be regarded as the destruction of the ordered conformations of the molecule into the disordered conformations. Cross-linking of biological macromolecules and loss of base pairing specificity may be regarded as the destruction of the orderliness of the cellular network. The reduction in the orderliness of the cell inevitably leads to the reduction in the functions of the cell. Therefore, the corollary points out that the destruction of the orderliness of the cell is the essential cause of cellular senescence.

Experimental results showed that generally speaking, if the cell can divide normally, then the cell will not start senescence. The phenomenon may be explained by the corollary of the hypothesis of orderliness. During the cell division, the damages accumulated inside the cell are distributed into two daughter cells and thus the cell division reduces the extent of the damages in the daughter cells by approximately half. If the cell keeps dividing and the speed of cell division is faster than that of damage accumulation, according to corollary 2, the cell will not begin senescence. In other words, if the effects of interference from the environment do not accumulate enough to interfere with cell division, then the cell will not start senescence.

> The dream of every cell, to become two cells.
> ——Francois Jacob

Chapter Four
The Principle of Cell Growth and Cell Division

4.1 Background

In order to survive, the cells have to grow and divide. Growth and division are an important part of cellular life. In fact, they are one of the most fundamental differences between the living system and the non-living system. The cell theory indicates that all the present cells come from the division of the existing cells. There are mainly two different types of cell division, mitosis and meiosis, which undertake different tasks and perform different functions for the survival of the organism. Generally speaking, somatic cells are involved in mitosis while germ cells in meiosis. Different types of cell division produce different results. Theoretically, after mitosis the two daughter cells have the identical genome and keep essentially the same structure and function as their mother cell. The result of mitosis is proliferation. The daughter cells produced in mitosis can continue to grow so that the total volume of the cells increases (the volume of two daughter cells in comparison with that of one mother cell). Comparatively, meiosis not only reduces the number of chromosomes of the daughter cells by half but also changes the contents of the chromosomes of the daughter cells in the processes of crossing over and independent assortment. The four daughter cells after meiosis usually do not continue to grow. The result of meiosis is the diversity of the genetic variations in the daughter cells but not the increase of the total volume of the cells. Obviously, the cells that are involved in mitosis and meiosis are endowed with different functions and display different characteristics and thus should follow different principles.

As far as the cell size concerns, eukaryotic cells are usually larger than prokaryotic cells, with the size of most eukaryotic cells between 10μm ~ 100μm and most prokaryotic cells between 1μm ~ 10μm. The size of the cells in multicellular organisms is generally between 20μm ~ 30μm. Obviously, neither plants nor animals adopt the strategy of increasing cell volume to achieve the growth of individuals. Instead, they employ the strategy of increasing cell number by continuous cell division. Somatic cell volume shows a simple periodic change: new daughter cells after mitosis have only half of the size of their mother cell in the beginning, but they can rapidly increase the cell volume to the same size as their mother by synthesizing protoplasm. Then another round of division begins (Xie, 2013).

In this chapter, two different kinds of cell division will be discussed and the mechanisms for the fundamental process will be investigated.

4.2 Relationship between Cell Division and the Axiom of Survival

At the first glance, there is some contradiction between the axiom of survival and cell division. During the cell division, a mother cell disappears and two daughter cells come into being. The contradiction lies in how to view the fate of the mother cell and the relationship between the mother cell and her daughter cells. In other words, this contradiction is related to the definition of life or the lifespan of the cell. Here I propose one solution to it. No matter in mitosis or meiosis, the division of a mother cell into two daughter cells may not be considered as the end of one life and the beginning of two new lives but as a continuation of one life, instead. Each divided daughter cell should be considered as part of the previous mother cell. The difference between mitosis and meiosis is that the daughter cells in mitosis have the same genetic make-up as their mother cell while those in meiosis have different genetic make-up from their mother cell. Therefore, when one of the daughter cells is dead, the life of the mother cell still continues if the other daughter cell is alive. More specifically, if cell A divides into two cell B's, and one cell B divides into two

Chapter Four The Principle of Cell Growth and Cell Division

cell C's and so on, we have a series of cells originating from the cell A. We may define all these cells as the "group A cells". Even when one cell in the "group A cells" is alive, life of the "group A cells" as a whole still continues. The lifespan of the "group A cells" should be counted from the birth of the cell A to the death of the last cell in the "group A cells". The lifespan thus defined may be termed specifically the group lifespan.

To the multicellular organisms like humans, the biological meaning of the group lifespan should be counted from the formation of the zygote to the death of the last cells in the organisms. To the unicellular organisms like bacteria, even though each divided cell is an independent entity, we may still define the group lifespan for the group cells originating from some certain cells as in the multicellular organisms. In this case, however, there is some arbitrariness in choosing the original cell for the group cells. But a hierarchical relationship does exist for these different groups. Even though they can live independently, I speculate that each cell originating from the same cell should be strongly associated with each other because they are part of the whole group, which is reinforced by phenomena such as bacterial colonies, bacterial biofilm (Nadell, et al., 2009) and bacterial quorum sensing (Higgins, et al., 2007). The phenomenon of such close association of the cells with the same origin may be termed the "cell entanglement", analogous to the mysterious phenomenon of quantum entanglement in which two microscopic particles from the same origin are so strongly associated that when one particle receives some interactions, the other one can feel instantaneously.

According to the above discussion, the relationship between cell division and survival may be proposed. If a cell does not divide, the damages or stresses from the environment may accumulate and cause the cell to age and die. The fate of survival of the cell is determined by only that cell. However, if the cell keeps dividing and forms a group of cells, then even when many cells in the group die, the organism as a whole still survive if only some of the cells in the group are alive. In this case, the fate of survival is determined by all the cells in the group. Clearly, the process of cell division provides better opportunity for organisms to survive. Therefore, cell division does not contradict with but support the axiom of survival.

4.3 Hypothesis of Mitosis and Cell Growth

In order to survive and keep internal homeostasis, the cell needs resources such as nutrients from the environment. Usually the cell absorbs resources more than the requirement for the survival and homeostasis. In fact, the cell absorbs extra resources for their growth and division. Growth and division are two fundamental processes of the cell. Cell growth is a highly regulated process and usually coupled with cell division, which is usually considered as the means to control the cell size. Much progress has been made in our understanding of these two processes and a lot of details about the mechanisms have been described (Schmidt, 2004). However, little is known about the driving force for these processes. Let us study cell growth and mitosis in detail in this section.

First, let us consider the following thought experiment. We put a live cell and a thermodynamic system into the same cultivation solution separately (Figure 4.1). Let us assume that the thermodynamic system has the identical membrane of the cell and contains the identical materials of the cell. The difference between the two systems is that the live cell has orderly structures inside the cell membrane and belongs to the living system while the thermodynamic system does not have any pre-assigned orderly structures and belongs to the non-living system. For the thermodynamic system, after sufficiently long period of time, the system is in equilibrium with the cultivation solution and stays at its most stable state and the size of the system will not change with time any more. Comparatively, the live cell grows and the size increases. After the size of the cell reaches some values, the cell will divide. The divided cells will grow and divide again. The process keeps going on and on until the population reaches saturation. If a tiny drop of such saturated solution (cultivation solution containing the saturated cells) is taken out and put into another identical cultivation solution, the cell will grow and divide again. The population of the cell will increase exponentially and will reach saturation again. The procedure can be repeated continuously until the Hayflick limit for the cell division (if the limit does exist) is met.

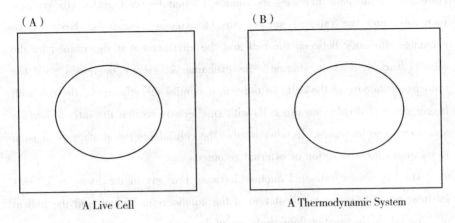

Figure 4.1　A Live Cell and A Thermodynamic System

Comparison of a live cell and a thermodynamic system in the identical cultivation solution. A. A live cell; B. A thermodynamic system with the same membrane and materials as the live cell.

The results of the above thought experiment clearly showed the fundamental difference between the living system and the non-living system. Even though the components of the two systems were the same, the conformations and structures of these components displayed tremendous difference in their behaviors and characteristics. This thought experiment illustrated the intrinsic properties of the cell during mitosis and gave some clues for the principles that the cell might follow in the process, as stated in the following hypothesis.

Hypothesis of Mitosis and Cell Growth: The somatic cell has the tendency to maximize the conversion of external resources into internal resources.

The resources here are defined as the nutrients and materials required for the synthesis of biological macromolecules such as nucleic acids, proteins, and also some other small molecules such as metal ions. According to their locations relative to the cell membrane, resources can be classified as external resources if they locate outside the membrane or internal resources if inside the membrane. The hypothesis indicates that the somatic cell has the intrinsic capability to do its utmost to convert the external resources into the internal resources spontaneously. To this end, the cell absorbs resources from the environment and consequently the cell increases its volume. When the cell grows, the volume increases, and the ratio of surface area to volume decreases. Because the exchange efficiency between the cell and the environment is determined by the ratio of surface area to volume, the exchange efficiency decreases with the increase of the size of the cell. In order to maximize the efficiency, the cell with bigger size will divide into two cells with smaller size so that the ratio of surface area to volume increases. In other words, the cell adopts the strategy of division to maximize the absorption of external resources.

The hypothesis indicated another intrinsic property of the living cell. This intrinsic property is the manifestation of the autonomy of biology and the holism of "the whole is greater than the sum of its parts". Clearly such intrinsic property belongs to the biological attribute of the living system and thus cannot be explained by the principles or laws of physical sciences.

It should be noted that the hypothesis only implies that the cell has the intrinsic tendency to maximize the conversion of the external resources into the internal resources. However, the tendency may not be necessarily fulfilled to its utmost because the fulfillment of the tendency depends on the impact of the environment which plays an important part in the fulfillment. For instance, the contact inhibition of proliferation prohibits cell division when there is not enough space or when there are not enough resources available in the environment. This impact is much more obvious when the cells are only the components of an organization or a structure in which the capability of division of the cells is most probably restricted by the functional or structural requirements of the system. For example, the cells forming human organs can only grow and divide to some extent, which should be ascribed to the functional or structural requirements of the human body. For unicellular organisms, this tendency is generally fulfilled to the most extent.

You may ask, "Why does the somatic cell convert external resources into internal resources?" I think the reason for this is that the internal resources can be utilized and manipulated by the cell for its own use while the external resources cannot and that the internal resources are the possessions of the cell but the external resources are not. The significance of the hypothesis is that the maximization of the conversion provides the cell with the maximal opportunity to survive. As discussed in section 4.2, the cell division gives rise to a series of cells forming the "group cells". The more cells in the group, the better chance for the group to survive. Therefore, when the cell maximizes the conversion of the external resources into the internal resources, it also maximizes its opportunity to survive. In this sense, the cell is selfish in possessing more its own resources for better chance of survival.

Cell growth and cell division are two closely associated processes. It is a long-lasting question of whether the growth leads to the division or the division leads to the growth. According to the hypothesis, the two processes work together for the same purpose of maximizing the conversion of the external resources into the internal resources more efficiently. Therefore the two processes may be considered as two different stages of one process. The purpose of the growth is for the division because the division produces more cells, which leads to better chance of survival. The division is for better growth because the divided cells absorb external resources more efficiently. It may be inferred that if the cell cannot divide, the cell will not grow continuously because the conversion efficiency decreases as the cell volume increases.

The above discussion may lead to the following corollary:

Corollary 3: The capability of division is the intrinsic property of the cell.

According to the hypothesis that the cell has the tendency to maximize the conversion of external resources into internal resources, the cell has to divide to fulfill this tendency. In other words, the cell has to divide spontaneously in order to absorb external resources more efficiently. Therefore the cell must be endowed with the capability to divide *per se*.

The terminally differentiated cells usually do not divide, which seems contradictory to the corollary. In my opinion, this phenomenon does not conflict with the corollary because the corollary only implies that the intrinsic capability of the cell. However, as discussed above, such capability may be restricted by the impact from the environment. In fact, the phenomenon only indicates that we do not observe the division of the terminally differentiated cells, which does not necessarily mean that the cells have lost their capability to divide. The fact that the cells do not divide may be due to the contact inhibition of proliferation imposed by the environment or other limitations caused by the structural or functional requirements of the organization that prevent them from dividing. Many of us may have the following experience in culturing cells. When the cell concentration closes to saturation, the population of cells does not increase significantly, which means that most of the cells do not divide. However, if we move a small part of the culture into another fresh cultivation solution, we find that the concentration will increase exponentially again soon, indicating that the depressed capability of division of the cell has recovered. Similarly, when the liver is surgically removed 70% of liver mass, the remnant tissue grows to recover the original mass and functions, implicating that the hepatocytes still retain the capability of division even though they do not divide when the liver has the normal size. These results may imply that the terminally differentiated cells still have the capability to divide, which will show up when conditions are permitted.

Corollary 4: The volume at which the cell divides may vary in a certain range.

For further discussion of the corollary, we may need to introduce some new concepts. The components of the cell may be classified as essential components and non-essential components. The essential components may be defined as the necessary components that are required for the cell to survive. Without any of these essential components, the cell cannot survive. Presumably the essential components of the cell have higher priority for synthesis in the cell growth.

The purpose of cell growth and division is to convert the external resources into the internal resources. However, at what volume or size of the cell does the cell divide? Considering from the hypothesis of mitosis and cell growth, the smaller the volume at which the cell divides, the higher the efficiency of the conversion. On the other hand, considering from the axiom of survival, the cell will divide only when the daughter cells can survive. In other words, the cell divides only after all the essential components of the cell have been synthesized. Therefore, these two factors determine the actual time or volume at which the cell divides.

Let us investigate this process in more detail. Essentially cell growth is a process of reproduction of cellular components such as nucleic acids and proteins while cell division is a process of distribution of these components. The cell exchanges resources with the environment through its surface and the exchange rate is determined by the ratio of surface area to volume. The rate decreases as the size of the cell increases. For the same volume, the sum of the surface areas of two small cells is greater than the surface area of one big cell (in the case of sphere, $S_2/S_1 \sim 1.26$, where S_2 is the total surface area of two small cells while S_1 is the surface area of one big cell). According to the hypothesis of mitosis and cell growth, a cell has the tendency to divide at small volume.

Then why do cells divide only after their volume or size reaches some certain values? The reason is that the cell obeys the axiom of survival. To survive, a cell needs to have all of the essential components of the cell, such as the important biological macromolecules and some other important components. If a cell divides before reproduction of all the essential components of the cell, at least one of the daughter cells must die. Based on the axiom of survival, this process should not occur. Cell division occurs only after reproduction of all the essential components of the cell. There was an experimental result in support of this idea: When a cultured amoeba was going to divide, a chunk of the cytoplasm was cut off. The amoeba then did not divide but continued to grow. If a piece of the cytoplasm was cut off again, then the amoeba did not divide. But if it is allowed to grow continuously, it would divide when the volume reached some size. Therefore, the cell divides only after all the essential components have been reproduced. In other words, the necessary condition for the cell to divide is that all the essential components of the cell have been reproduced, which gives the smallest volume for the cell division. However, this situation may not be the optimal conditions for the cell to divide because some auxiliary components in the cell may be more beneficial to better growth for the daughter cells, which may be determined by interactions between the environment of the cell. Different combinations of these auxiliary components may form a spectrum for the optimal conditions for the cell division. Therefore, the volume or the size at which the cell divides can varies in some range, with the lower limit being determined by the axiom of survival and the upper limit by the optimal conditions for the cell division.

4.4 Hypothesis of Meiosis and Cellular Genetic Diversity

Mitosis and meiosis are the two main different forms of division for eucaryotic cells. In mitosis, somatic cells grow and divide to produce more cells, with the genomes of all the cells being essentially the same. The divided cells still keep the capability of division. In meiosis, germ cells change not only their ploidy in chromosomes such as from diploid to haploid in humans but also their contents in chromosomes. Usually the resultant gametes in animals and many plants do not involve in mitotic division. For instance, the mature sperm and egg cells in humans do not involve in mitotic division.

Meiosis is an important mechanism for the species of organisms to survive in the changing environment. It produces many genetic variations, of which the inferior will be eliminated based on the principle of natural selection proposed by Darwin. The genetic variations come from two different processes in meiosis I: crossing over and independent assortment, with the former occurring first and the latter second. In the process of crossing over, the exchange of DNA segments between homologous chromosomes produces novel versions of chromosomes, which are different from the parental ones. In the process of independent assortment, homologous chromosomes appear randomly in the equational plate and migrate randomly to the two poles of the cell under the pulling of spindles. For human beings, independent assortment of 23 pairs of chromosomes yields 2^{23}, or more than 8 million, possible chromosome complements for every cell. Even though this process does not create new versions of genes, it provides the offspring with myriad novel combinations of chromosomes. For providing more genetic variations, the sex chromosomes X and Y act like short homologous chromosomes during meiosis. The total effect of the above two processes gives rise to a huge number of different genetic variations that provide genetic diversity for eukaryotic populations to adapt for changing environmental conditions.

Chromosomes of gametes after meiosis are genetically unique and have various combinations of DNA that are derived from both parents.

In most organisms, germ cells come from the first asymmetric division of the fertilized egg. Therefore, the reproductive differentiation during embryonic development has been decided in the earliest stage. The components of germ plasma are distributed unevenly in the posterior pole of the unfertilized egg cell. With the first asymmetric division of the fertilized egg, the daughter cell that contains the germ plasma will differentiate into germline stem cell, also known as the primordial germ cell. Germ cells produce offspring and transfer the genetic information to offspring through sexual reproduction. In the process of gametogenesis, the primordial germ cell produces gametes through sex determination and differentiation. In this process, the cells undergo meiosis, in which the chromosomes pair, recombine and change the genotype of the daughter cells. Spermatogonia and oogonia produce mature sperm and eggs by meiosis, respectively. In the process of spermatogenesis, four sperm are produced from a spermatogonium after two meiotic divisions. Although these four sperm are similar morphologically, their genotypes are different. Because sperm cannot proliferate by mitotic division and all sperm can only come from meiosis from spermatogonia, their genotypes are all unique. Similarly, a mature egg and three polar bodies are produced after two meiotic divisions from an oogonium and all eggs have different genotypes. This pattern of producing germ cells ensures the diversity in genotypes of the gametes.

The above discussion leads to the following hypothesis for the mechanism for meiosis:

Hypothesis of Cellular Genetic Diversity: The germ cell has the tendency to maximize its diversity in genetic variations.

The hypothesis indicates that the main function of the germ cell is to provide genetic variations for the next generations but not for the absorption of resources from the environment. In order for the species to have better chance to survive in the changing environment, the germ cell has the tendency to maximize the combinations of its genetic makeup, which may be selected during the process of fertilization. Therefore, meiosis provides a mechanism for the species to survive in the changing environment. Thus, from the view point of philosophy, this characteristic of the germ cell is consistent with the requirement of natural selection in the evolutionary theory established by Darwin.

The selection of high quality sperm is carried out during the process of fertilization. Take mammals for an example. The number of mammalian sperm in one ejaculation is up to hundreds of millions. These sperm must swim across vagina, uterus and cervix, and arrive at the site of fertilization through the fallopian tubes. They also need to cross the cumulus cells and corona radiata surrounding the egg and further go through the structure known as the zona pellucida to fertilize an egg. Only a few sperm can pass through the corona radiata, and usually only one of them can penetrate the zona pellucida and fertilize the egg. During their journey to fertilization, hundreds of thousands of sperm have to race in competition with their counterparts and conquer a lot of extremes such as the strong acid conditions in the female reproductive tract. Most of the sperm die and are dissolved and absorbed inside the body of the female. This mechanism of sperm selection inside the mother's body ensures that no inferior sperm can fertilize the egg. Obviously, selection of sperm in fertilization process plays a crucial role in producing high quality of offspring. This is indeed a vivid miniature of natural selection of survival of the fittest in the evolutionary process proposed by Darwin.

Because of their different roles in fertilization, sperm and eggs behave quite differently in meiosis. After puberty, sperms are produced hundreds of millions daily while eggs are arrested in metaphase of meiosis II, which can only be triggered to continue meiosis by fertilization of sperm. To produce high quality of offspring, two requirements must be met: 1) as many as possible genetic variations of sperm are available to be selected and 2) the egg cell itself is of high quality. In order to provide as many as possible genetic variations to be selected by eggs, sperm are produced in such a way that all the sperm have different genotypes, as discussed above. In other words, to guarantee the maximization of genetic diversity, mature sperm are not involved in mitotic division because mitosis produces all the daughter cells with the same genotype as their mother cell. Because the egg cells do not have the similar selection mechanism as sperm to prevent the inferior egg cells from fertilization, they are produced quite differently from sperm. Oogonia undergo asymmetric meiotic division and only one oocyte is produced with three small and short-lived polar bodies. This ovum gets all the resources (cytoplasm, mitochondria) from the oogonium. In this way, eggs with high quality are produced.

In summary, in order to produce high quality offspring to survive in the changing environment, the germ cell has the intrinsic property to maximize its genetic variations and the selection mechanism guarantees that no inferior gametes will be involved in fertilization.

4.5 The Economic Principle of Cell Division

As discussed in the previous sections, the living cell has the tendency to divide whenever possible. However, in some cases, the cell does not always divide, such as the nerve cells and the egg cells. This phenomenon may be due to the following principle of the cell division:

The Economic Principle of Cell Division: The cell will not divide if the divided cells do not perform any functions or do not contribute anything to biological processes.

The economic principle indicates that the cell divides only when necessary. Even though the cell has the intrinsic capability of division, namely, dividing when possible, the cell division is restricted by its environment and organizations, especially when it is only part of an organization. In other words, the cell divides only when its divided daughter cells are allowed by the environment or beneficial to the function and activities of the organism. The principle works as a regulatory mechanism that restricts the intrinsic capability of division of the cell and thus can be regarded as a negative feedback for the process of cell division. For the cells of unicellular organisms, the principle usually does not give much restriction to the cell division because every cell exists as an independent entity and thus its division is mostly only confined by the impact of the environment. Comparatively, the cell division in multicellular organisms is affected strongly by the functions that the cell is endowed with. For instance, in multicellular organisms, each kind of cells has its own task to undertake, which is usually fulfilled in the form of a tissue or an organ. Each tissue or organ has its optimal size to perform the function. When the cell number reaches some certain values and more cells will not enhance the function of the tissue or organ, the cell will not continue to divide. This phenomenon results from the economic principle of cell division.

Contact inhibition of proliferation (CIP) is a phenomenon in which the cell growth is arrested when cells have contact with each other. Recent research indicated that the phenomenon is initiated by the cell-cell adhesion receptors (McClatchey & Yap, 2012). In the process of culturing bacteria, the population of the cell increases exponentially at the beginning and then reaches saturation even nutrition is still abundantly available. CIP for unicellular organisms may be due to space limitation that prevents the further divided cells from surviving successfully. According to the axiom of survival, the cell will not divide in this situation. Comparatively, during the processes of histogenesis or organogenesis of multicellular organisms, the cells stop dividing when the tissues or organs are properly formed, which may result from the economic principle of cell division.

The economic principle of cell division clearly shows that even though the cell is endowed with the intrinsic capability of division, the capability is tightly regulated, controlled and sometimes may be severely suppressed by the environment especially when the cell is only part of an organization. Cells in different tissues or organs perform different functions and accordingly behave differently in cell division. For instance, the cells divide early and frequently if they age early and are damaged easily in their life such as the blood cell. In these cases, the cells divide to replace those cells that cannot function properly in the organization. On the other hand, cell division may not be easily observed in the organization in which the cells last long and are not easily damaged such as some neurons in the central nervous system. Therefore, according to the economic principle, some neurons do not need to proliferate. The reason why egg cells do not continue meiosis II until fertilization is because the function of the egg cells is to produce offspring and obviously continuation of meiosis II without fertilization does not perform this function. According to the economic principle, the egg cells do not continue meiosis. These phenomena emphasize that the cells divide only when necessary.

The economic principle of cell division may also explain some behaviors of stem cells in organisms with hierarchical organizations. Research indicates that almost all organs have corresponding stem cells, such as liver stem cells, neuronal stem cells. These stem cells usually remain dormant under normal physiological conditions. According to the economic principle, it may be speculated that when the terminally differentiated cells are damaged and need replenishment, the stem cells still remain dormant as long as the same kind of terminally differentiated cells can proliferate. The stem cells take part in mitosis and proliferate only when these differentiated cells cannot fulfill the task. This is also true for the cancer stem cells. The cancer stem cells usually remain dormant and only participate in mitotic division and differentiation when most of the cancer cells are killed by chemotherapy. This may be the difficulty for effective treatment of cancers.

What I cannot create, I do not understand.

———Richard P. Feynman

Chapter Five
The Principle of Cell Differentiation

5.1 Background

Multicellular organisms are usually made up of many different types of cells. Different types of cells in the same organism are distinct in structure and protein composition. They perform diverse functions in biological processes. However, most of these different types of cells have essentially the same genome and are usually originated from the same cell, such as the zygote, by means of cell division and cell differentiation. Cell differentiation is a process in which a cell changes from one cell type to another. The process dramatically changes the cell's characteristics, such as the cell size, shape and metabolic activities. Cell differentiation occurs in both multicellular organisms and unicellular organisms. Cells in multicellular organisms differentiate to form different tissues and organs with specific structures and functions while the cells in unicellular organisms differentiate to adopt different living styles to adapt to the changes in environmental conditions. Here, only the cell differentiation in multicellular organisms is studied.

5.2 Genome as the Identification of the Somatic Cell

Different types of somatic cells in an organism, such as blood cells and nerve cells, are quite different in morphology and phenotype. How to classify these different types of somatic cells? Considering the fact that all the different types of somatic cells in the same organism originate from the same zygote by means of cell division and cell differentiation, they should have essentially the same genome. Different types of somatic cells in an organism are only different in their gene expression profiles in the process of cell differentiation. Comparatively, the somatic cells in different individuals have different genomes. For instance, human chromosomes vary among individuals, averaging about one difference (polymorphism) per 1000 base pairs. It was estimated that the genome has variant versions at about 10^6 sites (Wolpert, 2009). Therefore the genome may be used to identify the somatic cells in one individual organism from the other. In other words, all the somatic cells in one individual organism can be regarded as the same entity with different conformations or in different states, while the somatic cells in different individual organisms are different entities. The above idea is supported by the fact that the immune system rejects stem cells taken from someone else even a close relative because these stem cells are regarded as foreign to the recipients (Wolpert, 2009).

The genome is the inherent identification of the entity of a somatic cell. A specific set of gene (namely, the luxury genes) expression profile represents a specific phenotype of the cell, or a specific state of the cell. The differentiation process may be regarded as a process in which the divided cells change their conformation from one state to another. For instance, ontogeny is a process in which the divided cells change their conformation from totipotent state to terminally differentiated state.

5.3 The Differentiation Potency and Differentiation Potential of the Somatic Cell

Multicellular organisms are usually composed of cells at different hierarchical levels, such as stem cell, progenitor cell and functional cell. Cells at different levels have different differentiation capability. For instance, a zygote has the highest capability while the functional cells have the lowest capability. Furthermore, stem cells have been classified as totipotent (zygotes), pluripotent (e.g. ESCs), multipotent (e.g. HSCs) and so on. To better describe such characteristic of the cell, a new concept termed differentiation potency needs to be introduced. Obviously, the more cell types a cell can differentiate, the higher its differentiation potency. Therefore, the differentiation potency may be defined as the number of different cell types a cell can differentiate. If a cell can differentiate into N different types of cells, then its differentiation potency is defined as N.

If we investigate the process in more detail, differentiation can be further classified into direct differentiation and indirect differentiation. The direct differentiation means that a cell directly differentiates into another different type of cell. The indirect differentiation means that a differentiated cell differentiates further into another different type of cells. Thus the number of cell types into which a cell can differentiate is defined as the sum of the number of the directly differentiated cell types and all the indirect differentiated cell types at various levels. Figure 5.1 gives a simple example. As shown in the figure, Cell A differentiates into two different types B_1 and B_2, which belong to the direct differentiation of A. B_1 differentiates into three different types C_1, C_2 and C_3 while B_2 differentiates C_4 and C_5, respectively. C_1, C_2, C_3, C_4 and C_5 are all assumed to be terminally differentiated cells. The differentiations of B_1 to C_1, C_2 and C_3 are the direct differentiation of B_1 and the differentiations of B_2 to C_4 and C_5 are the direct differentiation of B_2. All these differentiations are the indirect differentiation of A. According to the above definition, the differentiation potency for A is 7, B_1 is 3, B_2 is 2 and all Cs are 0, respectively.

During differentiation, the genome of the cell does not change but the expression of some specific groups of genes does change. In fact, differentiation is the result of the expression of different specific groups of genes in the genome. The different expression of the genome results from the different conformations of the chromatin structure (Ho & Crabtree, 2010; Orkin & Hochedlinger, 2011). For instance, the experimental results showed that the chromatin in the stem cells has the "open" conformation while the chromatin in the somatic cells has the "close" conformation (Meshorer & Mattout, 2010; Gaspar-Maia, et al., 2011).

To better describe the transition of the cell between different states, another new concept termed differentiation potential (hereafter referred to as potential) needs to be introduced. The potential is a characteristic of the differentiation state and is presumably associated with the conformations of the chromatin in the nucleus. The difference of the potentials between two states is an indicator for

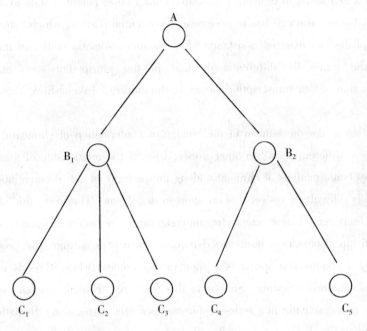

Figure 5.1 A Schematic Diagram of Differentiation Potency Showing the Potency of the Cell at Different States

how the transition may occur between the two states. If a cell can spontaneously transmit from one state to another, it implies that the potential of the initial state is higher than that of the final state. For instance, in the ontogeny process a zygote divides and differentiates to the terminally differentiated functional cells. This result indicates that the potential of the zygote is higher than that of the functional cells.

However, it is not always true that a cell can transmit spontaneously from a state with higher potential to any other states with lower potentials. More specifically, different states of potential are associated with different conformations of chromatin. The transition between different states of potential corresponds to the change of the conformations of the chromatin. If the change of the conformation is prohibited by some factors, say topological restraints, the transition is not allowed even though the initial state has higher potential than the final state.

If a cell wants to transmit (or jump) from a lower potential state to a higher potential state, the cell has to overcome the potential barrier, which can only be accomplished by external assistance. For instance, somatic cells can transform from the terminally differentiated state to the pluripotent stem state by introduction of four transcription factors in the culture (Takahashi & Yamanaka, 2006).

What is the mechanism of the change of conformation of chromatin in the process of differentiation? In other words, what is the relationship of a series of different conformations of chromatin along the pathway of cell differentiation?

It is difficult to answer this question in detail now. However, the following analog may give some clues to understanding the essential points of the relationship. Consider a harmonic dynamic system of a spring. The process of relaxing a compressed spring is a spontaneous process while it needs external force to compress a spring. Similar to this harmonic dynamic system, we may assume that chromatin in a series of different potential states along the pathway of differentiation has topologically homologous configurations; the more "compressed" configuration corresponds to the higher potential state and the less "compressed" configuration corresponds to the lower potential state. Therefore, differentiation is a process in which the chromatin changes from the more "compressed" configurations to the less "compressed" configurations. Definitely, further research is needed to unravel the real mechanism of the transition of the cell between different potential states in the process of differentiation.

The differentiation potency and the differentiation potential are two related but different concepts. The differentiation potency indicates how many different cell types a cell can differentiate while the differentiation potential is a characteristic quantity associated with the conformation of chromatin and the different potential of two states shows the possible ways of the transition between two different states of the cell.

5.4 Hypothesis of Cell Differentiation

Cell differentiation gives rise to different types of cells for various functions

Chapter Five The Principle of Cell Differentiation

in multicelluar organisms. For example, in vertebrates (including mammals) and humans, a fertilized egg differentiates into more than 200 different types of cells. The differentiation process is spontaneous and therefore the capability of differentiation should be the intrinsic property of the cell. So what is the driving force for this process? According to the discussion in the previous sections, I propose a hypothesis to explain the mechanism for this driving force.

Hypothesis of Cell Differentiation: The somatic cell has the tendency to minimize its differentiation potency.

The hypothesis indicates that the cell has the intrinsic property to decrease its differentiation potency spontaneously. Because cell differentiation is a process which is closely associated with the environment, the behavior we observe in some processes may be different from the prediction of the hypothesis. These phenomena are probably due to the interaction with the neighboring cells or intervention from the environment, which suppresses this intrinsic property of the cell.

Generally speaking, different types of cells along the pathway of differentiation have different differentiation potency. For instance, embryonic stem cell and hematopoietic stem cell have different differentiation potency. According to the hypothesis, all the cells have the tendency to decrease their differentiation potency. In other words, if you want to prevent the cell from differentiating, you must provide some special micro-environments such as stem cell niches to keep the cell in the same differentiation potency state. Therefore, cells at every state need appropriate micro-environment to maintain in the same state except for those at the terminally differentiated states. Based on experimental observations, it may be inferred that the capability of such micro-environment to accommodate cells is very limited. For example, the mammalian morula cells at the 8 cell stage, with the same differentiation potency as the zygote, can differentiate and develop into a complete individual under certain conditions. In other words, the micro-environment, or the stem cell niches, of the mammals can only accommodate 8 totipotent cells. If there are already 8 cells and the cells still keep dividing, the stem cell niches will not have enough facilities to hold the extra cells. In this situation, the extra cells will differentiate into the states with lower differentiation potency. In this way, the hierarchical levels of the differentiation potency of the cell are formed. This hierarchical structure of the cell may be analogous to the atomic structure of the energy levels of electrons.

The process of ontogeny in humans (and other mammals) may be explained with the hypothesis of mitosis and cell growth and the hypothesis of cell differentiation. Experimental results showed that most tissues in multicellular organisms have the corresponding stem cells and the number of these stem cells is very small. It can be inferred that there are only a few stem cell niches available for accommodation of the stem cells in the tissues. When the fertilized egg divides, the divided cells will first fill the stem cell niches for the totipotent cells. These divided cells keep dividing and the resultant cells will occupy all the stem cell niches for the totipotent states. The extra divided cells from further divisions will differentiate into the states with the next lower differentiation potency. Now the cells at this potency state may come from two different sources: 1) differentiation from the upper potency level; 2) division from the same potency level. It seems reasonable to assume that the second source provides more cells than the first one because the cells from the second source only undergo one division process while the cells from the first source have to undergo two processes: one division and one differentiation. When all the states at this potency level are occupied, the division of the cell at the upper level will stop. The reason is that the divided cells cannot differentiate into the states of the next lower level and according to the economic principle of cell division, the cell will not divide in this situation. The further divided cells from the same level will differentiate into the cells with the next potency level.

The same procedures can go on and on and, therefore, there will be a series of stem cells at different potency levels and also progenitor cells in the organism. The progenitor cells further differentiate into terminally differentiated functional cells. These functional cells do not differentiate any further because their differentiation potency is zero. In principle, when the terminally differentiated states are fully filled, the whole series of cells along the differentiation pathway will stop division and the ontogeny is complete. In

Chapter Five The Principle of Cell Differentiation

practice, however, this situation may rarely occur because the organisms are dynamic in nature and the cells are changing continuously. Presumably, the information of the number of cell types and the cell number of each type in an organism is contained in the genome of the cell.

During embryogenesis, the cells gradually change their differentiation potency from totipotency to pluripotency and finally to unipotency. For instance, the blastomeres before the formation of blastocyst in amphibian animals have the same differentiation potency as the fertilized eggs. These cells are totipotent and can differentiate and develop complete individuals under certain conditions. After formation of the three germ layers, due to the differences in the locations and the micro-environment of the cell, the differentiation capability of the cells is restricted to differentiate only into the cells of specific tissues and organs of the corresponding germ layer. These cells are pluripotent. After organogenesis, the fate of all the cells in the tissue is determined, namely, they can only differentiate into the specific type of cells. The cells now are unipotent.

Along with the decrease in differentiation potency of zygotes, the chromatin of the cell changes accordingly. During this process, different sets of genes are expressed. Are there any changes in the nucleus of the cell in this process? The experimental results of the nucleus transplant techniques indicated that the nucleus of the differentiated somatic cells retains the whole set of genes for developing normal individuals and the cells have the capability to develop the whole organism. Furthermore, the structure of the chromatin within the nucleus of the somatic cells can recover to the structure of totipotency in the cytoplasm of the egg cell. In other words, the changes in the chromatin structure in the process of differentiation of fertilized eggs can be recovered. In this sense, the differentiation process of a fertilized egg is reversible for chromatin structure, namely, under some conditions, the chromatin structure of the terminally differentiated cells can be recovered to the structure that has the same functions as that of the fertilized egg.

In addition, experimental results showed that by adding some appropriate transcription factors, the terminally differentiated cells can restore to the state of embryonic stem cell. For instance, with the help of retrovirus vector, the genes of the four transcription factors (Oct3/4, Sox2, c-Myc, Klf4) were injected into the mouse skin fibroblasts and the fibroblasts from the adult mice regained the pluripotency of embryonic stem cell (Takahashi & Yamanaka, 2006).

5.5 Relationship of Differentiationality and Functionality

Function and the degree of differentiation are closely related in the process of cell differentiation. The more a cell differentiates, the more specific its function is determined. In order to study further the relationship between them, the concepts of functionality and differentiationality of the cell need to be introduced. Functionality may be defined as percentage of the capability of a cell to fulfill its functions in actual biochemical processes. The function of a terminally differentiated cell is completely determined and hence its functionality is the highest and may be taken as 1; the function of a totipotent stem cell, on the other hand, is totally undetermined, namely not having differentiated towards any functional cells yet and thus its functionality is the lowest and may be taken as 0. Differentiationality may be defined as percentage of the capability of a cell to differentiate into all the different types of cells along a specific differentiation pathway. Let us assume that a totipotent stem cell undergoes N different types of cells to become a terminally differentiated cell. If a cell at a state in which it can differentiate into n different types of cells to reach the terminally differentiated state along the same differentiation pathway, then the differentiationality of the cell at that state is defined as n/N. Obviously differentiationality is 1 for the totipotent stem cell and 0 for the terminally differentiated cell.

Analysis of the experimental results shows that there may be some quantitative relationship between differentiationality and functionality of the cell. When the functionality of a cell is high, its differentiationality is low. On the contrary, when the functionality of a cell is low, its differentiationality is high. For example, the differentiationality of a totipotent stem cell is the highest and its functionality is the lowest. Comparatively, the differentiationality of a terminally differentiated cell is the lowest while its functionality is the highest. In the spontaneous differentiation process, the differentiationality of the cell decreases and its functionality increases. In the process of induced pluripotent stem cell, the differentiationality of the cell increases while its functionality decreases. If F = functionality, D = differentiationality, then in the process of the differentiation and dedifferentiation, the following formula should hold:

$$F + D = 1 \qquad (5-1)$$

Therefore, the sum of differentiationality and functionality of a cell is conserved in the process of cell differentiation and dedifferentiation.

The relationship of differentiationality and functionality of the cell can be described schematically by the differentiationality-functionality diagram, as shown in Figure 5.2. A state of a cell can be represented by a point in the diagram and all the states of the cell in the processes of differentiation and dedifferentiation are described by a line. For instance, the totipotent state can be represented by the point (1, 0), where the first number represents differentiationality and the second one functionality; the terminally differentiated state can be represented by a point (0, 1). Obviously, all the other states should be represented by the points in the line between these two states. For instance, in the process of differentiation, if a cell stays at such a state that it differentiates into N/2 different cell types to reach the terminally differentiated state, then the differentiationality is 0.5 (where N is the number of different cell types for a totipotent cell to reach the terminally differentiated state). At this state, it is reasonable to assume that the functionality of the cell is also 0.5. Therefore the diagram clearly demonstrates the differentiation process. The process of ontogeny is represented by a line moving downward from (1, 0) to (0, 1) while the process of dedifferentiation is shown by a moving upward line. In contrast, cell division is expressed only by a point in the diagram because there is no change in either the differentiationality or functionality in the process of division.

The "ground state" in physical sciences refers to the lowest energy state or the most stable state of a system. How to define the "ground state" for the living system? Which quantity should be used to determine the "ground state"? From the above discussion, differentiation potency may be an appropriate candidate for this purpose. For instance, a cell at the totipotent state keeps dividing and differentiating spontaneously in the process of ontogeny. During this process, the differentiation potency keeps decreasing. When they reach the terminally differentiated functional state, the cells cannot differentiate any further and accordingly the differentiation potency cannot decrease any further. This situation resembles the gravitational potential which decreases when an object on the earth goes down from higher places to lower places. Therefore, the terminally differentiated state may be regarded as the "ground state" for the cell in the process of differentiation, represented by (0, 1) in the differentiationality-functionality diagram.

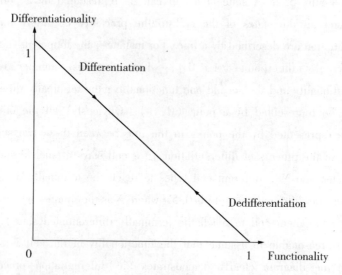

Figure 5.2 The differentiationality-functionality diagram showing the changes of differentiationality and functionality of a cell in the process of differentiation and de-differentiation

5.6 The Tree Structure of the Somatic Cells in Multicellular Organisms

As discussed in section 5.2, all the somatic cells in a multicellular organism may be regarded as representatives of one entity with different conformations or in different states. The hierarchical relationship of these cells can be classified according to their differentiation potency. Tracing the development of an organism, we can construct a diagram to show the hierarchical relationship of the cells in an organism, which may be called the tree structure of the cells in an organism. Figure 5.3 shows a schematic diagram of such a structure. As shown in the figure, the growth of the tree starts with the original cells such as a zygote or morula cells in mammals. These cells divide and differentiate into pluripotent embryonic stem cells. The tree then branches out to different types of multipotent adult stem cells. The adult stem cells can keep dividing and differentiating until the terminally differentiated functional cells are formed, as discussed in section 5.4. In this structure, the cells with totipotency are regarded as in the root state and the functional cells in the leaf states. The cells between these two states are regarded as in the stem states, which range from pluripotent cells to unipotent cells. According to how many times of differentiation to reach the stem state from the totipotent root state, the stem states can be further classified as primary, secondary and tertiary stem states. Therefore, the tree structure clearly demonstrates the position of the cells in hierarchy of the organism based solely on their differentiation potency.

The tree structure provides a clear picture for the functions that the cells perform in an organism. The terminally differentiated cells in the leaf states participate in all the detailed and actual biological functions of the organism such as the nerve cells and blood cells while the cells in the stem states work as the housekeeper in the organism such as to replace the damaged, aged or dead cells. Based on the tree structure, a hierarchical mechanism for repairing damaged, aged or dead cells may be proposed. When the cells in the leaf state are damaged, aged, or dead and need to be replaced, the cells in the same leaf state will divide to replace them if they can fulfill the task. If they cannot, the cells in the stem state next to the leaf state will undergo division and differentiation to replace them. Similarly, if some of the cells in the stem state need to be replaced and the other cells in the same stem state cannot fulfill the requirement, the cells in the nearest upper level of the stem state will divide and differentiate to do the job. Obviously, such a hierarchical repair mechanism makes the best use of the resources in the organisms and maintains the organism

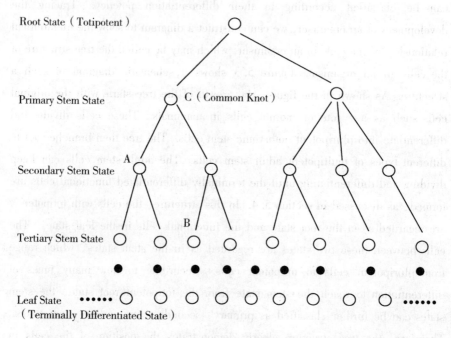

Figure 5.3 The Tree structure of the Cells in a Multicellular Organism

most efficiently. Such mechanism is supported by the fact that stem cells such as HSCs are quiescent, which are estimated to divide only five times per lifetime in mouse and most of the repopulation activity is accomplished by the committed progenitors, which is the downstream of the stem cell in the differentiation cascade (Zipori, 2009).

5.7 Transition between Different Differentiation Potency States

The cells can change their types in the processes of differentiation, de-differentiation and trans-differentiation. For instance, during embryogenesis, a zygote changes from the totipotent state to the functional state by means of cell division and differentiation. On the other hand, experimental results showed that the differentiation potency of differentiated functional cells can be increased by reprogramming (Hanna, et al., 2008), somatic cell nuclear transfer (SCNT)

(Jaenisch & Young, 2008), by adding some "core" transcription factors such as Oct-4 and Sox-2 (Takahashi & Yamanaka, 2006) or some small-molecule compounds (Hou, et al., 2013). These results clearly demonstrated that cells can change their differentiation potency in the processes of differentiation or dedifferentiation. Such change of the cell in the differentiation potency states can be regarded as the transition of the cell between these states.

In the process of trans-differentiation, a cell can change from one type to another in different differentiation pathway. Based on the tree structure, a possible mechanism for trans-differentiation may be proposed. If a cell wants to change from one type to another in different differentiation pathway, it must first de-differentiate to at least the nearest higher potency state that are common to both initial and final pathways, called the common knot, as shown in Figure 5.3, and then re-differentiates from that common knot to the final state. For instance, if the cell in the A state wants to transmit to the B state, the cell needs to de-differentiate from the A state to the C state and then re-differentiate from the C state to the B state (Figure 5.3). Further investigation is needed to discover the real mechanism for the process of trans-differentiation.

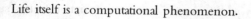

> Life itself is a computational phenomenon.
>
> ——Claus Emmeche

Chapter Six
A Conjecture of the "Final Rules" of Theoretical Biology

6.1 Background

Life is a magical phenomenon in the nature. She exists ubiquitously in our materialized world, bringing us rich and splendid scenes. One of the main goals of biological research is to understand the essence and connotations of life. We human beings started the journey of exploring the meaning of life from the time of Aristotle and pursued unremittingly the truth of life in human history. We made great efforts and have made some achievements already, but we still have not understood completely the essence and connotations of life. When breakthroughs are made in science, we usually expect that these scientific breakthroughs can help us understand better the nature and meaning of life. However, the results are not always satisfactory. There are many reasons for this situation, some of which may be the limitations of our knowledge and techniques.

As stated by Richard P. Feynman, "What I cannot create, I do not understand". In order to understand life, scientists have been working continuously to synthesize an artificial life which is the system that is synthesized from pure chemical components and is capable of Darwinian evolution, with the hope of unveiling the mechanisms and principles of life. Steven A. Benner pointed out that whether or not we can synthesize such a system could be regarded as a criterion for the validity of a theory of life. He described vividly that "If life is nothing more than a self-sustaining chemical system capable of Darwinian evolution and if we truly understand how chemistry might support evolution, then we should be able to synthesize an artificial chemical system

Chapter Six A Conjecture of the "Final Rules" of Theoretical Biology

capable of Darwinian evolution. If we succeed, the theories that supported our success will be shown to be empowering... In contrast, if we fail to get an artificial life form after an effort to create a chemical system... we must conclude that our theory of life is missing something" (Benner, 2009).

One of the most important features of life is self-replication, which is certainly the first characteristic targeted by synthesizing artificial life. The history of scientific development showed that the first work of synthetic life began in computer. In the 1930s, Alan Turing, the father of theoretical computer science and artificial intelligence, invented a general purpose computer, known as the general Turing machine. In principle, the machine can be used to compute any computable sequences written on a tape. In the 1940s, John von Neumann developed Turing's ideas and constructed a self-reproducing machine. By means of a series of logical manipulation, the machine could create a copy of itself which could produce another copy of itself. Since the double helical structure of DNA molecule was determined in 1953, the biological role of DNA in genetics has been established. The development of molecular biology has advanced our understanding of the fundamental processes of the living system at molecular level. DNA is the carrier of the genetic information and the genome of the living cell contains all the information for the cell. Therefore, the DNA sequence is of great importance for us to understand the essence of life.

In recent decades, the technique of genome sequencing has been developed by leaps and bounds. Today, genome sequences of many species have been determined, including the human beings. As the technique has been improved, it will be possible to sequence the genomes of any species in which we are interested. One of the most important questions that we may ask is how much information and how reliable the information can we obtain from the sequence of genome. In other words, how can we understand the essence of life from the available genome sequences? This is indeed a critical question we need to answer. In this chapter, I will analyze the similarity between the machine code of computer programming language and the genome of the living cell and try to answer the question of what information we can extract from the genome of the living cell.

6.2 Artificial Intelligence

Recently artificial intelligence (AI) attracted most human attention again. In March 2016, AlphaGo, a computer program for the game of Go, beat the professional world champion Lee Sedol in a five-game match with a final score of 4 to 1. This was the first time in human history for a computer program to beat a world professional champion in the game of Go. The game of Go is usually considered as the most complicated classic game for human beings and the most difficult classic game for computers to conquer our human beings. But AlphaGo made it now! The design of AlphaGo is different from that of the previous Go programs in that it plays the game with two different neural networks called "brains", one of which uses the "value network" to estimate the situation and the other uses the "strategy network" to choose the move (Silver, et al., 2016). The main principle of the technique "deep learning" adopted by AlphaGo is similar to that of the biological neuronal brain: forming the "brain" neural network to process precision calculation and the network can be further strengthened through examples and experience.

One of main goals of developing AI is to make a machine be capable of "thinking" as human beings. Machine thinking is actually the calculation of the computer. Generally speaking, the computer mainly does three things: receiving input, doing calculation and sending output. Regardless of the contents of input and output, the computer regards them as nothing but a pile of data. In order to make a machine be able to think through computer programs, scientists have been investigating intelligent algorithms which can correctly and effectively obtain the optimized results for actual situations by calculation of all possibilities based on input data. Theoretically, by using the artificial intelligent algorithms, the machine can do any repetitive work done by human beings or even those that cannot be fulfilled by human beings.

In order for AI to be able to simulate the thinking process of human beings, scientists and engineers studied the process of human thinking? The human brain is a complex neural network which is composed of neurons. Each neuron seems very simple, just receiving electrical signals from some neurons and sending out

Chapter Six A Conjecture of the "Final Rules" of Theoretical Biology

electrical signals to stimulate other neurons. Even though the working mechanism of neurons is very simple, neurons can perform their functions intelligently when they have enough in number to form the neural network with connections among all of them. For example, the human brain works intelligently because it contains about 100 billion neurons and on average each neuron connects with other neurons with about 7000 synapses.

The research of artificial neural network has a long history and dated back to 1943 when W. S. McCulloch and W. Pitts established a mathematical model for neurons, called the MP model. With the help of the model, they showed that single neurons could perform logical functions, thus opening the era of artificial neural network research. In the 1960s, Widrow proposed a self-adaptive linear network, based on which non-linear multiple layer self-adaptive network was developed. In the 1980s, Hopfield established a model for neural network and Linsker put forward a new theory of self-organization for the perceptron network. In recent years, artificial neural network has been mostly developed towards simulation of human cognition, combining with the fuzzy system, genetic algorithm and evolution mechanism to become computational intelligence. Furthermore, introduction of convolution, an important concept in neuroscience, into artificial neural network has produced tremendous results in many fields, such as speech recognition and image classification.

Artificial neural network is a mathematical model for information processing, which is similar to the structure of synaptic connections in the human brain. It consists of a large number of nodes (or neurons), which can receive input signals from "neurons" of the upper layer and assign different weights according to the importance of the "neurons". Then the "neurons" sum up the weighted input signals and integrate the result into a function. After calculation, the final result will be output to the "neurons" of the next layer in the neural network. By using a series of sophisticated algorithms and being trained by a large amount of data, the artificial neural network can work similarly as the neural network in the human brain. It can figure out the "features" from the complicated data and can produce the results that require smart thinking.

Then how does the artificial neural network learn? The process of learning in artificial neural network is essentially the adjustment of weights of each "neuron" so that the whole artificial neural network can perform well enough to meet the requirement of testing tasks. Take the task of classification for an example. First, each element in a learning set is input to the artificial neural network and the network is trained how to classify. After completion of the whole learning set, the network will summarize its own ideas and rules based on the experience it has obtained from the examples it has learned. Then, we will test the network with the examples in a testing set. If the network passes the test (e. g. 90% correct rate), then it has been constructed successfully and it can be used to handle the transaction of classification.

Artificial neural network has the ability of self-adaption and self-organization. It can change the synaptic weights in the learning or training process in order to adapt to the requirements of the environment. It is also a system with learning capability, which will develop its own knowledge so that it can possess more than the knowledge that is input originally by its designers. Usually, there are two ways of training. One is the learning under supervision, in which classification or imitation is performed on the given standard samples. The other is the learning without supervision, in which only the ways of learning or the rules are given. The specific contents of learning vary with the environment (i. e. the input signals) and the system can automatically find the characteristics and regularities. This way of training is more similar to the human way. For instance, two different methods of training were used for the programs of the game of Go. AlphaGo was is trained with supervision while AlphaGo Zero was trained without supervision. AlphaGo Zero could reach the level of professional players relying completely on its own ability of learning, without any help from researchers or input data. The results showed that AlphaGo Zero performed much better than AlphaGo (Silver, 2017). Therefore, AI can now "learn" and "think" in the human way.

The ideas and techniques used in AlphaGo and Alphago Zero may be also applied to other AI domains that require long-term planning and decision-making, such as the problems of protein folding, reduction of energy consumption and searching for new materials. These achievements indicate the powerful future of AI. However, in principle, no matter how smart AI will be, all of its functions and activities are determined by the computer programs designed and completed by the human beings. The programs are first written in the form of some high-level programming languages such as C++, Java, Python, Prolog and Lisp, which are termed source codes of the program. Then these source codes are compiled into the long strings of various combinations of the basic elements of 0 and 1, which are termed machine codes of the program. AI can behave according to the information and directives retrieved from these machine codes. Obviously, different levels of source code designs lead to different levels of behaviors of AI, such as the simple mechanical actions in the 1980s and the complex neural network "thinking" of AlphaGo and AlphaGo Zero in recent years. Therefore, the fundamental changes in the algorithms will change the ways of "thinking" and features of behaviors of AI. In other words, the revolution in computer technology and algorithm can totally change the intelligence level of AI. This process may be referred to as the evolution of AI. The excellent performance of AI such as AlphaGo and AlphaGo Zero convinced us that the simple mechanism of artificial neural network can result in the sophisticated behaviors and exquisite thinking like human beings.

6.3   Some Features of the Genome of the Living Cell

The cell is the smallest structural and functional unit of all organisms. Most of the biological processes of living systems are carried out inside the cell. As stated by E. B. Wilson that all the answers to biology are eventually sought in the cells because all organisms are or once were a cell, the fundamental problems in biology are ultimately the problems of the cell. Structurally, eukaryotic cells are composed of cell membrane, cytoplasm and nucleus, with their chromosomal DNA being stored inside the nucleus. Each of the three components plays its role in the cell, cooperating with each other to maintain the normal functions of the cell. By comparison, prokaryotic cells do not have nucleus and the chromosomal DNA is stored in the area of nucleoid.

Schrödinger was the first to investigate the basic concepts and principles of biology from the perspective of physical science. He considered that the genetic behaviors of the biological systems are determined by the "aperiodic solid" — the chromosome fiber (Schrödinger, 1944). James Watson and Francis Crick proposed the double-helix model for DNA molecules based on the X-ray diffraction pattern studied by Rosalind Franklin, implicating the biological role of DNA in genetics (Watson & Crick, 1953). Further development of molecular biology has improved our understanding of the essential role of genes in biological systems.

It is commonly accepted that the cell can perform all the functions and fulfill all the tasks in the right order and at the right time according to the information retrieved from the genome. Scientific investigation implied that the genome contains all the information for the cell. More specifically, at molecular level, the central dogma indicates that the genetic information flows from DNA to RNA to protein. Almost all the features and characteristics of the cell are represented by proteins and the functions of the cell are mostly performed by proteins. The genome determines which proteins are synthesized and at what time the proteins are synthesized and regulated. At the cellular level, the genome directs the behaviors of the cell in the biological processes such as cell growth, cell division and cell differentiation.

Much progress has been made in our understanding of the behaviors of genome in the cell. Research in developmental biology indicated that the cytoplasm of oocytes could directly or indirectly interact with the genome and change the expression pattern of the genes, which is called cytoplasmic memory. For instance, the cultured kidney cell nuclei of the frog *Xenopus laevis* were injected into the enucleated oocytes of the newt *Pleurodeles waltlii* and the results showed that the resultant cells did not synthesize the proteins that were usually synthesized in the kidney cells but only synthesized those that were normally synthesized in the oocytes (De Robertis & Gurdon, 1977). Therefore, the cytoplasm of an oocyte can activate some originally inactive genes and inactivate some originally active genes of the somatic cells, implicating that the conformation of the genome is affected by the impacts from the environment.

Systematic studies with the technique of the somatic cell nuclear transfer (SCNT) demonstrated that the cytoplasm of an oocyte can change the expression pattern of the genome. SCNT is the process by which the nucleus of an oocyte is substituted with the nucleus of a somatic cell such as a skin cell or a liver cell. The enucleated oocyte and the somatic nucleus are fused by inserting the somatic nucleus into the "empty" oocyte. The host oocyte can reprogram the somatic cell nucleus and the resultant cell behaves as an embryonic stem cell and the activities and functions of the cell are dominated by genome of the donated somatic cell (Wilmut, et al., 2002). The results showed that the somatic cell nucleus differentiated from a zygote retains all the information necessary for development of an organism. In other words, the structural change of chromatin during the process of differentiation is reversible in some sense: as long as the conditions are appropriate, the chromatin structure in the somatic cells can recover to the state of the embryonic stem cell. For instance, a research team led by Ian Wilmut successfully cloned the sheep "Dolly" with the nuclear transfer method from a cultured cell line.

The recent work of J. Craig Venter and his research team clearly demonstrated that DNA was the software of life. Venter and his colleagues synthesized chemically the entire genome of *M. mycoides* and transplanted the synthetic genome into an *M. capricolum* cell (Lartigue, et al., 2007). They showed that the synthetic genome dominated over and replaced the original genome of the wild-type *M. capricolum* cell and the recipient cell became an *M. mycoides* cell. This was the first time in human history to transplant a totally chemically synthesized genome (the naked DNA) into a bacterial cell to produce a living cell. Their work indicated that the totally chemically synthesized genome can function as the same as the natural genome in the living cell and therefore implied that the living cell may be controlled by the digital information from a computer. Venter summarized in his book that, "The assumption, at least by most molecular biologists, was that DNA and the genome, represented by the sequence of letters in the computer, was the information system of life. Now we had closed up the loop by starting with the digital information in a computer and, by using only that information, chemically synthesized and assembled an entire bacterial genome, which was transplanted into a recipient cell, resulting in a new cell controlled only by the synthetic genome" (Venter, 2013). Most recently, a digital-to-biological converter was designed and completed (Boles, et al., 2017). It can receive digital information from a computer and convert into biopolymers, such as DNA, RNA and proteins, marking a great step forward towards the synthesizing an artificial life.

The above discussion clearly demonstrated the important role of DNA in the living cell. If we replace the genome of an organism, we substitute the species of the organism. Although the above result is only the replacement of a life, which is still not the creation of a life in the strict sense, the meaning is still far-reaching. The characteristics of life, including its species, are uniquely determined by the sequence of the genome.

6.4 The "Final Rules" of Theoretical Biology

In the previous sections of 6.2 and 6.3, we had discussions about the capabilities of AI and the features of the genome of the living cell. The advancement of technology of computer science is astonishing as demonstrated by the evolution of AI from simple mechanical actions to complex logical thinking and self-learning as human beings. On the other hand, the complex and exquisite ways of human behaviors may originate from the information contained in the genome, a long string of combinations of four nucleotides of A, G, C, and T.

Indeed, it is of great importance to notice the fantastic similarity between the genome of the cell and the machine code of AI. Genome has four elements of A, G, C, and T while the machine code has two elements of 0 and 1. We believe that the genome contains all the information for the cell and the machine code has all the information for AI. AI responds to stimuli according to the instructions and directives retrieved from the machine code and the cell presumably does so in the similar way.

Based on perspective of physical sciences, Schrödinger speculated that chromosomes must contain "some kind of code-script determining the entire pattern of the individual's future development". He deduced that the code-script had to contain "a well-ordered association of atoms, endowed with sufficient resistivity to keep its order permanently". He also explained how the

Chapter Six A Conjecture of the "Final Rules" of Theoretical Biology

number of atoms in an "aperiodic solid" could carry sufficient information for heredity. He argued that this solid did not have to be extremely complex to hold a vast number of permutations and could be as basic as a binary code, such as Morse code (Schrödinger, 1944).

Watson and Crick discovered the double helical structure of DNA and demonstrated the specific way of the code-script to transfer the genetic information from generation to generation. In the double helical structure, DNA molecules are composed of two strands and each strand runs in the opposite direction. The nucleotides in one strand are complementary to those in the other strand: adenine (A) base pairing with thymine (T) and guanine (G) base pairing with cytosine (C). Therefore, each chain in DNA can serve as the template for reconstructing the other. In this way, the information contained in the genome can be copied and transmitted to the offspring.

In the 1970s, Frederick Sanger developed the method of DNA sequencing with chain-terminating inhibitors and determined the sequence of the first full DNA genome for bacteriophage φX174 (Sanger, et al., 1977). In 2001, the first complete sequence of the human genome was determined and we had our first real view of the remarkable details of the aperiodic solid that contained the code for human life.

Indeed, we have made great advancement in genome sequencing. Even though we believe that the genome of a living cell contains all the information for the cell, such as cell growth, division and differentiation, we can only retrieve very limited information from the sequence itself. The coding region usually occupies about 2% of the genome. Even though some non-coding regions in the genome may play some roles in regulation of gene expression, organization of chromosome architecture and control of epigenetic inheritance, there are still many regions in the genome that we do not understand their biological functions and roles.

The task of computer scientists is to design the most smart and efficient AI by optimizing the architectures and strategies in the programs according to the grammar (or the rules) of the programming language. Because AI does not understand these high-level languages, the programs written in the programming language, or the source codes of the programs, have to be compiled into the machine codes so that AI can understand the instructions and perform the functions accordingly (Figure 6.1A).

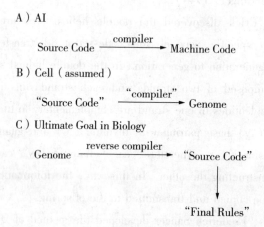

Figure 6.1 A Schematic Diagram of the "Final Rules"

Analogous to AI, the cell might only understand the genome and can carry out the functions according to the information contained in the genome. Therefore, it may be postulated that the genome of the cell is the "machine code" which comes from compilation of the programs written in the "biological programming language" for the cell (Figure 6.1B). The grammars or the rules of the programming language may be considered as the "Final Rules" of theoretical biology (Figure 6.1C). Here, I propose a conjecture of the "Final Rules" for theoretical biology:

Conjecture of the "Final Rules": The "Final Rules" are the principles of "biological programming language", which may be obtained by reverse compiling of the genome of the cell.

According to the conjecture, the ultimate goal for biological scientists is to decipher the genome of the cell and to discover these "Final Rules". In fact, the research in molecular biology has made some progress towards discovering the "Final Rules". For instance, we have already found some related elements for the biological programming language, such as the "start" and "stop" codons in the genetic code, which indicate the beginning and ending of the process of protein translation. In addition, the recent research in non-coding DNA sequences indicated that these regions of DNA play important roles in regulating gene expression. Epigenetic modification provides another way of modulating the expression of the genes. These results demonstrated the roles of the DNA segments in regulating the behaviors and properties of the cell.

Certainly there is still quite a long way to go to achieve this goal. There are so many hurdles in front of us and we are still not clear how to overcome them. The technique of DNA sequencing has been advanced tremendously and the accuracy has been improved a lot. Now, we have the ability to sequence any genomes in which we are interested. However, even if we have the sequences of all genomes at hand, it does not necessarily mean that we understand the essence of life. Some scientists pointed out that we have the sequences of all human genes but we cannot understand them.

As discussed above, the genome may be only the "machine code" of the "biological programming language". If this is true, it would be very difficult for us to understand the behaviors and characteristics of the living system from the sequence of genome because the information we get from the sequence of genome may be ambiguous and even misleading. Therefore, we should keep it in mind when we explain the sequence of genome that the genome may be only the "machine code" of the programs.

The road is tough but I am still optimistic about the future. Even though it is not clear now where exactly we should go, the current knowledge may provide some clues for future search for the "Final Rules". The information contained in the sequences of genome can be revealed during the process of gene expression. Experimental results showed that gene expression is associated with not only the sequences of the DNA fragments but also the locations of these DNA fragments in the three-dimensional structures of the whole DNA molecule, as illustrated by different patterns of gene expression in different types of cells in the same organism. More specifically, we all know that the genomes in eukaryotic cells are packed with histones inside the nucleus and the cells with the same genome in multicellular organisms may adopt different cell types, from stem cells to different terminally differentiated cells, indicating that they undergo different cell differentiation pathways and have different patterns of gene expression. It may be assumed that the genome of eukaryotic cells contains information not only in the sequence of the DNA but also in the steric arrangement of the DNA fragments. In other words, different steric arrangements of the same DNA fragments may represent different information. Obviously the information stored in this way is much more effective than the way in which the information is only stored in the DNA sequence. Therefore, the different steric arrangements of DNA fragments in eukaryotic genome may represent different statement in the biological programming language, maybe analogous to the "pointer" in the programming language of computer.

Comparatively, the genomes of bacteria are packed differently from those in eukaryotic cells. In bacteria, genetic material is not enclosed in a membrane-enclosed nucleus but has access to ribosomes in the cytoplasm and therefore the processes of RNA transcription and protein translation can occur at the same time. The genome of bacteria is essentially "naked" DNA, a two-stranded circular DNA molecule, and the mRNA after transcription do not have introns and can be directly translated into polypeptides. Also DNA replication occurs at both direction of the circular molecule. In addition, bacteria adopt different strategies of methylation of DNA from eukaryotic organisms to protect their own DNA from dissecting by their restriction enzymes. For instance, some of the genes cloned in eukaryotes are not recognized by the expression system of prokaryotes. All these differences in packaging of genomes and the detailed gene expression in prokaryotic and eukaryotic cells suggest that the genomes of the two different kinds of cells may come from compilation of the "source code" written in different "algorithms" of the biological programming language.

It is expected that molecular biology and other branches of biological sciences will contribute greatly to our understanding of the meaning of the genome and enrich our knowledge of the biological programming language. When we obtain the "Final Rules", we will have full understanding of the complete picture of the biological world. We will know the fundamental mechanisms of all the biological processes of the living system, such as the mysterious process of ontogenesis in human body. For instance, the formation of various organs in different locations in the body during ontogenesis may be determined by the statements like "go to" and "proliferate when". All this information is presumably contained in the previously thought "junk DNA" regions of the genome.

The position of the "Final Rules" in biology may be similar to that of the Newton's Laws in classical physics. The "Final Rules" can determine the behaviors and characteristics of the living system. It may be expected that the axiom of survival and the four basic hypotheses for the living cells discussed in the previous chapters, which belong to the intrinsic properties of the living cells, should be contained in the "Final Rules". Also the laws that the cells should obey when they interact with the external environment, which are the topics for further theoretical research in biology, should also be contained in the "Final Rules".

6.5 Implications of the "Final Rules"

Definitely, the genome of the living system contains all the information, which is enough to guide the behaviors and to determine characteristics of the system. However, how can we correctly dig out the information from the genome to understand the principles that govern the living system? From the comparison between the AI and the cell, we know that the DNA sequence of genome may be only the "machine code" of the cell. Therefore, understanding of the principles of the living system at the level of genome is analogous to understanding of the principle of AI at the machine code level. Even though all the characteristics and behaviors of AI can be determined by the machine code, which comes from compilation of the high-level programming language, it will be extraordinarily difficult to understand the features and principles of AI from the machine code, a long string of 1's and 0's. Similarly, although the genome of the living cell is enough to determine the characteristics and behaviors of the cell, it is also extraordinarily difficult to understand the principles of the cell from the genome, the "machine code" of life. Therefore, in order to understand the principles and laws of life, we need to decipher the "machine code" to obtain the "source code" and the "grammars" of the biological programming language. Only at the level of "source code" can we understand completely the living system and the essence of life.

Honestly, the idea of "Final Rules" is preliminary and now is more conceptual than practical. However, it may provide a new way of explaining the inconsistency between some theoretical predictions and experimental results and also some observations that cannot be explained by existing theories. In addition, it will provide a new direction for exploring life.

As discussed in section 6.1, to synthesize an artificial life is an important means to understand the essentials of life. The first and the most important step for such a grand project is to synthesize a living cell, which will open a door to our understanding of life. A definition of a synthetic cell is given by Church and Regis in their fantastic book, "A truly synthetic cell is one that we create ourselves, from the ground up. This could be a new form of living matter fabricated out of pure ingredients" (Church & Regis, 2012). Venter and his research team made a gigantic step forward to such goal. Their results demonstrated that to bacteria the totally chemically synthetic genome completely controlled the host cell as the same as the natural genome and the genome was the software of life (Venter, 2013). They also found in their experiments that one single base-pair deletion in more than one million base pairs could make the difference of life and no life (Venter, 2013). This finding implied the difficulty of changing the traits of an organism at the level of genome. The present search for modifying traits of an organism is still essentially hit-or-miss. This situation can be explained if we take genome as the "machine code" of an organism. If the concept of the "Final Rules" is correct, the functions and the roles of the cell are determined by the programs written with the biological programming language. It is easy to understand how difficult and how uncertain to modify the characteristics and behaviors of AI by changing the machine code of the programs. Similarly, we can imagine how difficult to understand the biological programs at the "machine code" level and how uncertain to modify the traits of an organism by means of changing the nucleotides in the genome.

The "Final Rules" will not only help us understand the fundamental processes of the living system but also can provide an effective and direct method to modify the traits of the living system. In the recent decades, synthetic biology has made great progress. By redesigning and recoding the genomes, synthetic biologists have been successfully modified microbes to have many different specific functions, such as producing fuels, detecting arsenic in drinking water (Church & Regis, 2012). Once obtaining the "Final Rules", we will be able to make tremendous contributions to the synthetic biology. The "Final Rules" will make biosynthesis much easier and more effective. For instance, we may realize the proposal made by Church and Regis "… to make human beings immune to all viruses, known or unknown, natural or artificial" (Church & Regis, 2012). Furthermore, we may use the biological programming language to write a program for *de novo* synthesis of a living cell, which is indeed a synthetic cell in the real sense. We may expect that the "Final Rules" will give us a comprehensive and thorough understanding of the biological world.

中文翻译

所有生物学的答案最终都要到细胞中寻找,因为所有生命体都是或者曾经是一个细胞。

——威尔逊(EB Wilson)

第一章
生物学理论研究的基本问题

1.1 背景知识

生命是什么?自从亚里士多德时代以来,许多哲学家和科学家一直在寻找这个看似简单问题的答案,但到目前为止,我们还没有一个明确而令人满意的答案。埃尔温·薛定谔(Erwin Schrödinger)在 20 世纪 40 年代提出了这个问题,并试图在物理学的框架内回答它。他认为染色体是一种非周期性固体,其中每一组原子都起独立作用,并提出了一些概念如维持生命系统的负熵流(Schrödinger,1944)。恩斯特·迈尔(Ernst Mayr)从哲学的角度来探讨生命的意义,认为所有的生物过程不仅受自然法则的控制,也受到遗传程式的控制(Mayr,1997;2004)。换句话说,他认为所有生命体都服从两种因果关系。一种因果关系是自然规律,加上偶然性,完全掌控着精确科学世界中发生的一切。另一个因果关系由独特的生物世界的遗传程序组成。现代生物学的这种二元论在本质上是物理化学的,它源于生物兼有基因型和表型的事实(Mayr,1997;2004)。斯图尔特·考夫曼(Stuart Kauffman)从整体论的角度来研究生命,认为生命是一种突现现象,当分子多样性超过某一复杂性的阈值时就会出现生命(Kauffman,1995),因此,生命是相互作用的分子系统的总体性质。阿迪·普洛丝(Addy Pross)从系统化学的基础来研究生命系统。他通过引入动态动力学稳定性(dynamic kinetic stability)的概念,声称生物学可以整合到化学中去(Pross,2012)。Marcello Barbieri 在《有机密码》(*The Organic Codes*)这本书中列出了在不同领域工作的科学家提出的超过 60 种典型的生命定义（Barbieri，2003）。这些定义中有许多是如此不同，以至于我们很难想象他们所描述的是同一种东西。

确实，我们生活的世界是如此多姿多彩和复杂深奥，以至于在不同领域工作或有不同背景的科学家获得了对世界的不同印象和理解。上面所提到的对生命的定义就是一个例证。然而，自然是简单的。我认为如果科学理论不是简单的，那是因为科学研究还没有揭示自然的本质。生物学的理论探讨也许可以从对系统的划分开始。世界上的一切物质都可以分为生命系统或非生命系统。要进一步研究生命是什么以及生命系统的基本原理是什么，我们首先要阐明生命系统和非生命系统的定义。生命系统通常被定义为具有复制和新陈代谢特征的任何系统。在这本书和以后的章节中，生命系统被定义为任何有生命的系统；而非生命系统被定义为任何没有生命的系统，包括那些仅由生物大分子组成但没有生命的生物系统。我将在第二章给出生命的定义。

现在让我们研究生命系统和非生命系统之间的不同行为。我们看看河里的一条鱼和一截木头（如图1.1所示）。众所周知，木头沿着水流顺流而下。虽然河床地形可能是非常复杂，但原则上我们可以根据物理科学的原理和规律来描述这一木头的运动轨迹，并能预测其在未来任何时刻的位置。至于那条鱼，我们根本无法做到这一点。我们不知道下一刻这条鱼将怎样游和游向何方。我们无法用物理学的任何原理或定律来预测它的踪迹。同样，我们可以预测一个球总是滚下斜坡，但是我们却不知道在同一个斜坡上的蚂蚁将会怎样走和走向何方。我们周围有很多这样的例子，这些例子揭示了生命系统和非生命系统截然不同的行为，同时也显示了物理科学在描述生物体行为方面的局限性。

除了行为的不可预测外，生命系统和非生命系统的另一个显著区别是行为的不可重复性。物理学和科学哲学都认为观察事件的客观性和重复性是密切相关的。在实验上发现的事件必须是可重复的才能被科学所接受。但是，对于生命系统，客观性和重复性却并不是总是一致的。例如，在前面所提到的在斜坡上的蚂蚁，不同时间，同一只蚂蚁，它所走的轨迹很可能是不一样的。而不同时间的同一个球，其轨迹相同，都是沿着斜坡往下。这是生命系统和非生命系统的另一个根本区别。显然，我们不能因为蚂蚁的轨迹不能重复而否定蚂蚁轨迹的客观存在。

科学的目的就是找到一些定律来描述世界万物的行为和特征。科学研

图 1.1 鱼与木头

注：活鱼和木头的不同行为表明，生命系统和非生命系统在自然界中遵循不同的原理（Yu Peng 作）。

究始于非生命系统，因为它们比生命系统更容易理解，更简单且易研究。从伽利略到牛顿，再到相对论和量子力学理论的建立，对非生命系统的研究取得了巨大成就，这使我们想当然地认为世界上所有的一切，包括生命系统，都遵循着相同的物理学定律而且生命系统与非生命系统之间的差异对于我们了解世界无关紧要。确实，科学发展的历史表明，物理科学支配着科学的各个领域。科学哲学及其基本思想和基本概念主要是以物理科学的实验结果和理论分析为基础而建立和发展起来的。自从经典力学诞生以来，我们对宇宙和生命的认识就一直铭刻着物理科学的印记。我们已经习惯于这些思想和概念，以至于我们本能地接受任何与之一致的东西，并拒绝任何与它们不一致的东西。传统的科学哲学把物理学作为所有科学学科的标准或标准范式，从不把生物学当作一门独立的科学学科来对待。还原论者认为，生物世界里的所有现象最终都可以用物理科学的原理来解释。

史蒂芬·霍金（Stephen Hawking）和伦纳德·蒙洛迪诺（Leonard Mlodinow）指出，科学哲学已经远远落后于现代科学的发展。在《大设计》一书中，他们说："传统上，这些问题（比如'宇宙是如何运作？''什么是现实的本质？'）是哲学的问题，但是哲学已经死了。哲学没有跟上现代科学的发展，特别是物理学。"（Hawking & Mlodinow, 2010）在我看来，生物学的情况比物理学还要严重滞后得多。

自 20 世纪 80 年代以来，生物学的发展突飞猛进，尤其是分子生物学和干细胞生物学领域。正如达尔文所指出的，生物学的成就将促使哲学走向新的繁荣。确实，生物学的迅速发展催生了一门新的科学哲学——生物学哲学。新哲学强调，生物学应该摆脱传统科学哲学的束缚，建立自己的科学理论、概念和架构。生物哲学最基本的问题之一是生物学在自然科学中的位置。更具体地说，生物学应该是独立于物理科学框架的一门自主的科学学科，抑或是仅仅属于物理科学框架内的一个分支。这个问题对于理论生物学来说至关重要，因为它不仅决定了理论生物学的框架，而且决定了生物学发展的方向和途径。在众多生物学哲学的积极倡导者中，迈尔是最重要的人物之一，他深入研究生物学，特别是进化生物学，分析了生物学在科学中的作用，得出生物学是科学中的一门独立学科的结论（Mayr, 1996）。他说："我越来越清楚地看到，生物学是一门完全不同于自然科学的科学，它的主题、历史、方法和哲学都有根本的不同。"（Mayr, 1988）

生命系统是极其复杂的，我们观察到的现象往往是把系统的内禀性质和系统受到环境的干预而做出的反应交织混杂在一起。为了建立一个生物学的理论框架，我认为，我们首先需要从生物过程中的行为和活动来识别什么是生命系统的内禀特性，什么是生命系统的受激反应。只有区分了决定生命系统行为和活动的内外因素，才有可能以正确的方式去探索生物学理论。本书主要探讨生命系统的内禀特性。

1.2 生物学的自主性

生物学在科学中的地位有两种对立的学派：还原论和自主性。还原论者认为生命系统和非生命系统都是由相同的化学元素组成，因此，原则上生命系统应当遵循与非生命系统相同的原理和定律。他们相信生物学并不独立于物理学，而分子生物学的研究最终将把整个生物学还原为物理学，因此生物学仅仅是物理科学的一个分支。还原论者使用一些策略来反驳生物学自主性。例如，他们把生物学分为两部分，第一部分可以简化为物理科学，服从普遍规律；第二部分不能简化为物理科学，因而被认为是非科学的（Mayr, 1997）。因此，还原论者声称没有必要承认生物学的自主性。自 20 世纪 80 年代以来，随着生物学的巨大发展，科学家开始认识到物理学及其相关的经典哲学概念不足以解释生物科学中观察到的现象。生物学的发展导致了另一种生物学哲学学派的建立，即生物学的自主性，这与还原论是相对立的。自主性的支持者认为，生物学的研究对象、概念结构和方法论与自然科学完全不同。他们相信生物学不能被简化为物理科学，生命的原理应该建立在一个独立于物理科学的新框架之上（Mayr，1997）。

  还原论者认为生物学不具备非自主性的原因来源于他们对还原论的哲学信仰。还原论者认为，通过研究其最小组成部分原则上可以获得对系统的解释，而系统的知识将从其分解后的较小部分中得到（Mayr，2004）。当然，原则上一个系统可以通过研究它的较小组成部分而被理解。然而，为了达到这个目的，必须满足一些条件。例如，我们需要知道较小的组成部分形成系统时的相互作用机制。让我们对生命系统做一个思想实验。假设我们把一个生命系统分解成属于非生命系统的更小的组成部分。尽管我们完全了解这些较小的组成部分，但我们仍然不能了解整个生命系统的内禀特性，或者从较小的组成部分中获得整个生命系统的知识，因为我们还不知道从非生命系统到生命系统的转化机制。事实上，当我们把一个生命系统分解为更小的非生命系统时，我们已经"杀死"了生命系统。在这种情况下，生命系统失去了它的生命特性以及有关生命的信息，而这些特性和信息是无法从属于非生命系统的组成部分中重新组合出来。因此，还原论在生物学哲学中面临着巨大的挑战。

  现在是我们重新考虑物理科学的定律和原理对生命系统的适用性的时候了。考虑到生命系统的基本特征，物理科学的定律和原理，如能量最小化原理和熵最大化原理，在描述生命系统的特征和行为方面既不直接也不有效。例如，非生命系统最有利的状态就是最稳定的状态，如动力学系统的最低能量状态或热力学系统的最大熵状态。显然，非生命系统的最稳定的状态根本不是生命系统的有利状态，因为能量最低或熵最大的状态意味着生命的终结，因此生命系统应该避免这些状态。可见，生命系统和非生命系统在本质上具有不同的性质，因此在各自特有的过程中应遵循不同的定律和原理。我们对生命系统感兴趣的状态并不是最稳定的状态，而是生命系统具有功能最佳和活力最强的状态。生物科学的一个目的就是要寻找优化生命系统的功能和活力的条件，而不是寻找能达到最低能量或最大熵状态的条件。因此，物理概念和物理量，如稳定性、能量和熵，并不是描述生命系统的最佳和最有效的特征量。生物学在理论上应该有自己的概念和特征量。这些概念将直接来自于生物世界中的现象，并且独立于物理世界。这些自主性概念是理论中基本和特有的元素，代表着生命的内禀特征，因此不能也不应该用物理科学来解释。我们需要一套全新的理论框架来描述生命系统。

1.3 生命系统的属性二象性

让我们用前面提过的例子开始讨论：在河里的一根木头和一条鱼（如图1.1所示）。我们知道鱼会沿着我们无法预测的轨迹运动。那么一条死鱼呢？经验告诉我们，死鱼会跟河里的木头一样漂流。让我们拿活鱼和刚死的鱼作个比较。两条鱼的化学成分应该基本相同，但其行为完全不同，一个遵循物理学的定律而另一个则不遵循。显然，两者的唯一区别是一个有生命而另一个却没有生命。这种差异完全来自于生命的本质。正是生命的本质使得生命系统和非生命系统的行为完全不同。换言之，生命赋予生命系统一些非生命系统所不具备的特质。当生命一旦离开生命系统，这些特质就立即消失。从这个简单的例子得出一个深刻的结论：生命系统具有属性二象性这样一个内禀特性，一个是物理化学属性，另一个是生物属性。生命系统的物理化学属性遵循物理科学的定律和原理，生物属性遵循自主生物学的定律和原理。

这两种不同属性在生命系统的行为中起什么作用？让我们来研究另一个例子：一只活狐狸从悬崖上掉下去。在这种情况下，无论狐狸思考什么，想做些什么，地球的引力在其运动轨迹中起关键作用。结合对图1.1的讨论，我们得出结论，生命系统的属性二象性是确定其行为和轨迹的两个因素。哪个属性在行为和运动轨迹中起主导作用，取决于生命系统所处的环境。相比之下，非生命系统只具有物理化学属性，其行为完全受物理和化学定律和原理的支配。因此，非生命系统的行为可以用物理和化学来解释和预测。当生命系统失去生命时，它将成为一个非生命系统，此时只具有物理化学属性。

以上论述表明，生命系统具有属性二象性，一个代表非生命世界的特征，遵循物理科学的定律和原理，另一个则代表生命世界的特征，遵循独立于物理科学之外的定律和原则。因此，生物属性对于我们理解生命的意义起着重要的作用。

那么生物属性从何而来？首先，我们需要考察生命系统和非生命系统的结构特征。两者之间的巨大差别在于生命系统组织结构的高度复杂性，尤其是层次等级结构的复杂性。这种层次等级结构是由于在某个层次等级结构的组织之间的相互作用而形成了一个较高层次等级结构的组织。这些结构的复杂性赋予了生命系统特殊的性质和能力，如对外界刺激的反应，以及生长、增殖和分化的能力（Mayr, 2004）。一般来说，当部分组成整体时，就会产生新质。这种现象既存在于生命系统中，也存在于非生命系统中。整体论认为"整体大于其部分之和"，因此新质不能通过重构整体中各组成部分的知识而推导出来。例如，水分子的液态特性是从氢原子和氧原子的形成中产生的，但不能从氢原子和氧原子的特性推导出来（Mayr, 1997）。生物体构成了一个更加复杂而且具有层次等级结构的系统，从细胞、器官到整个生物体。每一个较高等级层次的组织结构的特性都不能从其较低等级层次的组分的知识中推导出来。显然，当组分属于非生命系统而整体属于生命系统时，其新质肯定会更加巨大和令人惊讶，例如由分子组成细胞这种情况。因此，由非生命系统组成生命系统所产生的新质应当被视为生命系统的内禀属性和生命系统的生物属性的来源。

鉴于生物系统具有属性二象性，即物理化学属性和生物属性，以及物理科学无法解释其生物属性，当物理科学的原理和定律不能解释生物系统的某些新质时，我们不应该感到惊讶。事实上，正是这些新质才是生物学应该主要研究的部分，它们使生物学独立于物理科学。

迈尔指出，如果能够在地图上绘制物理科学和生物科学的领域，就会发现有相当多的重叠区域。重叠区域代表了物理科学和生物科学共同的部分，而非重叠区域显示了生物和非生物之间不可调和的差别（Mayr, 1996）。我认为物理科学知识和生物科学知识存在着重叠和不重叠的区域在某种意义上证实了生命系统的属性二象性的存在。因此，生命系统的属性二象性符合物理科学和生物科学的知识结构。

迈尔也提出了生命系统在生物过程中的双重因果关系。他说，所有的生物过程都服从两个因果关系，一个是精确科学中的自然定律，而另一个是描述生命世界特征的遗传程序（Mayr, 1961; 2004）。他说："两个世界（生命世界和非生命世界）都遵循物理科学发现和分析的普遍定律，但是生物体还服从第二类因果关系，即遗传程序的指令。而第二类因果关系在非生命世界中是不存在的。"（Mayr, 1997）迈尔进一步解释说："他们的（生物体的）活动是由含有历史获得信息的遗传程序所控制，这在非生命世界中是不存在的……现代生物学的二元论始终是物理化学的，它源于生物体具有基因型和表型的事实。由核酸组成的基因型需要其来理解进化的解释；而根据基因型提供的信息构建的表型，包括蛋白质、脂类和其他大分子，需要功能性（近因的）解释来理解。这种二元论在非生命世界中是未知的。对基因型和表型的解释需要不同的理论。"（Mayr，1997）

从上述讨论可以清楚地看出，迈尔提出的双重因果关系与生命系统的近因和远因都有关联。双重因果关系中的其中一种属于表型的功能生物学范畴，另一种则属于基因型的进化生物学范畴。这两种因果关系在本质上都是物理化学的。相比之下，我在这里提出的生命系统的属性二象性与迈尔提出的双重因果关系是不同的。在我提出的生命系统的属性二象性中，其中一种属性是属于物理化学的，而另一种是属于生物学的。这两种属性都只涉及生命系统的近因即功能生物学范畴。正如前面所讨论的，二象性中的生物属性既不是超自然，也不是活力论，本质上它是由于属于非生命系统的组分之间的相互作用而在生命系统中产生的新质。

二象性的概念在科学中并不罕见。例如，波粒二象性是微观世界中粒子的特征。生命系统的属性二象性也许是生命世界的基本特征之一。

事实上，正是生命系统的生物属性使得生物学成为一门自主的学科。遗憾的是，目前大多数工作主要致力于研究生命系统的物理化学属性而不是生物特性。这可能出自三个主要原因：（1）认识不到生物属性的存在，认为物理化学属性是生命系统的唯一属性；（2）忽视生物属性的重要性；（3）因其"非客观"特性，将生物属性排除在科学研究的范围之外。

对生命系统的研究可以在个体有机体、器官、组织、细胞、分子和原子这些不同等级层次结构水平上进行。那么，我们也许会问："我们在不同等级层次水平上进行研究会得到什么样的信息？"在展开进一步讨论之前，我想先给出几个定义。物理学性质定义为描述系统状态的可观测量的任何性质，而化学性质是指在化学反应过程中或化学反应后材料变化明显的任何性质；也就是说，只有通过改变物质的化学特性才能建立的任何性质（维基百科）。我认为，生物学性质可定义为与生命所特有的过程和活动相关的任何性质。一般来说，我们从原子到个体有机体的各个等级层次水平的研究中都能获得系统的物理学性质，因为对于所有这些等级层次，存在诸如力、能量和稳定性的性质。然而，只有在分子和更高等级层次水平的研究才能得到系统的化学性质。系统的化学性质主要和分子在化学反应中的化学键的断裂和形成相关。显然，在原子水平的研究中无法得到系统的这些性质。同样地，系统的生物学性质只有在细胞和更高等级层次水平的研究才能得到。这是因为根据现有的细胞理论，细胞是生命最基本的结构单位和功能单位，那么在分子等级层次水平和低于分子等级层次水平的研究都不能得到与生命有关的直接信息。我认为，当我们在分子等级层次水平上来研究生命系统时，我们只是研究其物理化学属性而不是生物学属性，因为生物学属性的信息不会存在于任何低于细胞等级层次水平的结构中。

当然，生命系统的物理化学属性是非常重要的。对这种属性的研究在分子水平上揭示了生命系统的特征和原理，为我们了解生命打开了大门，为生物学研究奠定了基础。例如，沃森和克里克的DNA双螺旋结构帮助我们理解了遗传的奥秘。然而，我们应该意识到生命系统的另一个重要属性，即生物学属性。目前的生物学研究往往忽略了生命系统的这一基本属性，甚至把物理化学属性作为系统的唯一属性。更重要的是，我们也应该意识到目前生物学研究中所使用方法的局限性。例如，在分子生物学的研究中，我们不可避免地将细胞切开来研究细胞内部的分子。在这种情况下，细胞基本上失去了生命的信息，尽管其组成部分基本保持不变。因此，我们从分子生物学中得到的是细胞的物理化学属性，而不是细胞的生物学属性。因此，如果说"生命现象"可以用物理科学来解释，那么它本质上就是物理化学现象，或者说是生命的物理化学属性。Addy Pross 提出了描述生命系统的动态动力学平衡（dynamic kinetic equilibrium）概念，并声称生物学可以还原为化学（Pross，2012）。我认为他可能只是把生命系统的物理化学属性还原为化学，而不是把生命系统的所有属性都还原为化学。在此，我必须强调，上述讨论并不意味着在分子水平上对生命系统的研究是不恰当的或毫无意义的，而只是指出生命系统在不同层次上有不同的特征，而把所有这些不同层次的特征结合起来有助于我们全面地理解整体生命的现象。

1.4 细胞作为理论生物学研究的对象

生物世界中的生物体在功能体和结构上是多种多样的。在组织层次上，每个物种都有自己的特点。达尔文也许是成功地系统研究生物学普适理论的第一人。他所研究的基本单位是整个生物个体。他考查了由这些生物个体所组成的生物种群的性质并得出了物种演化的一些重要原则。由于当时科学方法和技术的明显限制，他没有研究比个体有机体更低组织层次的结构。随着20世纪80年代生物科学的发展，我们在细胞和分子水平上对生物体的过程和结构有了更深入的了解。知识的积累使我们有可能在这些水平上对生物体的原理和机制进行理论分析和探索。

什么结构层次用来研究生物学的普适理论才合适呢？从结构上来讲，生物圈可分为生物种群、物种、个体生物体、器官、组织、细胞和分子。我认为细胞是现阶段理论研究的最佳选择。原因很简单。在低于细胞组织的层次结构中，通常没有生命活动。尽管有些核酸和蛋白质复合物组成的病毒参与了某些生命活动，但它们无法独立生活，而且从它们无法建立起更高的层次组织。因此，它们不是理论研究的一个好的选择。在比细胞更高的组织层次中，不同物种的结构和功能差异很大，很难研究和分析它们的普遍行为和机制。作为理论生物学的起始点，我们首先要研究的问题是生命系统与非生命系统的根本区别及其不同的行为和特征。显然，细胞是实现这些目标的最佳选择，因为细胞具有生命系统的所有特征，而且能够反映出生命系统和非生命系统之间的根本区别。正如美国生物学家E. B. Wilson指出的那样，所有生物学的答案最终都要到细胞中去寻找，因为所有的生命体都是，或曾经是一个细胞。

我们也许会问在分子水平上对生命系统的描述是否比在细胞水平上的描述更为准确。我的回答是"不一定"。当然，分子水平上的描述具有更多的微观细节。但是，更详细的描述就一定是更准确吗？我的答案是"不一定"。这取决于我们所研究的对象。例如，用显微镜观察细胞会比用肉眼观察细胞更准确。然而，用显微镜观察足球就不会比用肉眼观察足球更准确。尽管用显微镜观察足球可以给我们更多的微观信息，但用肉眼观察能够让我们更好地了解足球的特性和功能。

至于细胞水平上的定律和原理能否最终还原到分子水平上的定律和原理的问题，我的回答是"不"。根据生物的自主性，这两个不同结构层次的定律和原理属于不同的世界，一个是生命的世界，另一个是非生命的世界，而和生命系统的生物学属性有关的那些定律和原理不能进一步还原，因此这个问题的答案是"不"。尽管从分子生物学中获得的机制和规则可能有助于理解生物学中的现象，但它们不应作为判断细胞水平理论是否有效的标准。细胞的内禀特性是所有成分之间相互作用的总效应，它们反映了整个系统的集体性质，因此也不是像生命力那样神秘的东西。

一个复杂的生物体可以被视为一个细胞社群，它通常是由数十亿个不同类型的细胞组成。这些细胞形态各异、功能不同。细胞主要参与五个不同的生物过程：生长、分裂、分化、衰老和死亡。这些过程和细胞的生命密切相关。通常，生长使得细胞的内部物质增多、体积增大，分裂导致细胞数量增加而分化增加了细胞的种类。衰老和死亡是两个密切相关的过程，衰老通常和死亡联系在一起。生物学的理论研究将探讨这些基本过程的原理和机制。

作为一个生命系统，细胞在结构上非常复杂，并且与环境有着广泛的相互作用。这些相互作用很强，在细胞的过程和活动中起着至关重要的作用。然而，作为理论研究的第一步，我们需要研究细胞的内禀特性和生命系统与非生命系统的根本区别。因此，在这本书中，我主要研究细胞作为一个孤立系统时所具有的特点和行为。而细胞之间的相互作用以及来自更高结构层次的组织，如器官和个体生命体的信号和影响将在以后的研究中讨论。虽然这是理想状态下的理想模型，但对孤立细胞的研究将能够揭示细胞的内禀特性，阐明生命系统与非生命系统的区别，因此能够为生物学理论的进一步研究奠定基础。

此外，为了专注于细胞内禀特性的研究，除非另有说明，书中的讨论总是假定细胞是处于理想的条件之下，即（1）足够的养分；（2）没有来自环境的任何干扰或损害。

器官的形成……不会是为了对主体造成痛苦或伤害。

——查尔斯·达尔文

第二章
生存与生物学的目的性

2.1 背景知识

自然界中的生命系统和非生命系统是两种完全不同的系统，具有截然不同的行为和特征。从本章到第五章，我们将考察这两种不同系统之间的根本区别。本章中，我们将研究两种系统之间最本质的区别：目的性。目的性在科学中是敏感的，因为传统的观念认为，目的性以及相关的术语如目的论等都属于主观性范畴，因而被排除在科学研究的范围之外。尽管我们观察到生命系统的行为和活动都普遍存在着目的性的特征，但在科学讨论中总是避免使用这个术语。就算被使用，也只被认为是描述生命系统现象的一种便捷手法。本章我们将从哲学和科学的角度，深入讨论目的性的概念和诸如生存这类相关的问题。

作为生命的基本结构单元和功能单元，细胞参与所有的生物活动。那么细胞心目中有没有一些活动比另外一些活动有更高的优先权？它的行为和活动遵循什么规律？在下面的章节中，我们将详细回答这些问题。

2.2 生存公理

生存公理：在现有环境下生存是细胞在其所有生命过程和活动中的终极目的。

这里所提出的公理适用于简单的单细胞生物体如细菌和酵母以及构成复杂的多细胞生物体如人类的细胞。这个公理表明，细胞以生存为其生物学过程和活动的最重要的准则。换句话说，如果任何过程或活动对细胞的生存不利，那么这些过程或活动将被终止或至少被修改，以便获得更好的生存机会。

这一公理得到了很多事实和观察的支持。例如，溶液里的细菌游向葡萄糖，但远离毒素，因为葡萄糖有利于它们的生存，而毒素则有害于它们的生存。为了生存，许多细菌通过表达具有抗生素耐药性的蛋白质来抵抗抗菌治疗，如体外泵或经过不断进化后产生出来的对常规抗生素产生耐药性；它们也可以通过改变自己的结构以避免被抗生素杀死。癌细胞经过代谢重组以维持生存和快速增殖。此外，癌症干细胞通过调节特定基因的表达来抵抗外界的压力，奋力生存。例如，通过增强跨膜 ATP 结合盒转运体的表达，癌症干细胞可以逆浓度梯度将化疗药物从细胞内转移到细胞外，从而降低药物在细胞内的浓度，避免化疗药物的杀伤作用。这是癌症多重耐药性的主要原因（Leonard, et al., 2003）。

一些基本的生物学过程也可以用这个公理来解释。例如，所有细胞必须在细胞分裂之前完成其 DNA 的复制。这就确保了每一个分裂的细胞都得到生存所必需的所有遗传物质。细胞需要控制分裂的时间和频率，以便正确地发育，得到更好的生存。显然，细胞的过程和活动总是以目的为导向的，都可以通过其为了在现存条件下生存这个目的来解释，如果不直接关联，则必须间接地作为一种手段来达到这一目的。

尽管生存的概念已被广泛应用于进化生物学，但是，将生存这个概念引入到功能生物学的理论研究中仍然是必不可少的，因为它揭示了细胞的基本特性，并指出生命系统和非生命系统的本质区别。

生存公理与两个基本概念有关：目的性和生存。由于目的性是一个非常复杂的问题，涉及科学和哲学的许多方面，我将在 2.4 节中讨论。在这里，我将首先讨论生存问题。

2.3　生存公理和细胞死亡的关系

我们确实观察到细胞死亡。应该强调，生存公理并不是说细胞不会死亡，而只是指细胞会尽一切可能避免死亡。细胞死亡主要有三种不同的形式：凋亡、自噬细胞死亡和坏死。细胞死亡也可分为被动死亡和主动死亡。被动死亡是指细胞受到不可挽回的损伤，因而被杀死，而主动死亡则表示细胞积极参与其死亡，实质上是自杀。大多数自杀是由细胞凋亡引起的（Green, 2011）。

在各种形式的死亡当中，凋亡和主动死亡似乎与生存公理相矛盾，因此需要更详细的讨论。根据这个公理，细胞在所有的过程和活动中都以生存为终极目的，因此不应参与凋亡或主动死亡。为了进一步讨论，需要对主动死亡和被动死亡做出更详细和具体的定义。我认为，主动死亡应该是细胞本身触发死亡途径的过程，而被动死亡应该是细胞外部的信号触发死亡途径的过程。根据这一定义，即使细胞参与了死亡的途径，细胞的死亡也不一定是主动死亡。正是触发信号的来源决定了细胞的死亡是主动的还是被动的。

细胞凋亡通常发生在包括动植物在内的多细胞生物体中，也可能发生在酵母等一些单细胞生物体中（Green，2011）。有一些对单细胞生物体中的程序性细胞死亡的研究将利他自杀归因于亲属/群体利益，但结果和解释不明确，还需要进一步研究（Nedelcu, et al., 2010；Green，2011）。因此，这里的讨论只集中对多细胞生物体的研究。

多细胞生物体通常由许多种不同的细胞组成，每一种细胞都发挥其特定的功能，以保持机体的完整性。例如，我们人类一共大约有200种不同的细胞，约10^{14}个细胞。生物体中的所有细胞都是协同配合的，因此生物体可以被看作是一个由细胞组成的社会。而且，生物世界中存在着一种普遍现象，就是一个生物体的上层组织结构具有抑制下层组织结构行为的能力。例如，人体的组织有能力抑制其组成成分的细胞的行为。当某些病变发生在细胞中时，细胞可能会尽一切可能修复这些损伤，从而获得更好的生存机会。有时细胞能成功修复，但有时会失败。在失败的情况下，这种细胞的存在可能不利于组织甚至整个生物体的生存，这时，组织或生物体将触发保护机制来消灭这个细胞。因此，正是更高等级的组织迫使细胞死亡，以获得自身更好的生存机会。生存公理对于具有不同等级层次结构的生物体具有等级层次的意义。

细胞凋亡是通过细胞内的某些途径发生的，这给人的印象是细胞主动参与了死亡过程，并被视为自杀的证据。细胞内的各种凋亡途径中，最常见的途径是线粒体外膜被破坏而触发凋亡的线粒体途径（Green，2011）。促凋亡的BCL-2效应物通过引起线粒体外膜通透（MOMP）而促进凋亡，而抗凋亡BCL-2蛋白则通过避免线粒体外膜通透（MOMP）而阻止凋亡。BH3-only蛋白在这一过程中起着重要的作用，因为它们可以通过调节这两类BCL-2分子来决定凋亡是否发生。BH3-only蛋白通过激活促凋亡的BCL-2效应物或通过降低抗凋亡BCL-2蛋白的活性而引起凋亡。因此，BH3-only蛋白在决定细胞命运中起着关键作用。BH3-only蛋白在不同的环境条件下表达和调控不同。因此，它们是在细胞凋亡中信号传输的靶点，也可被视为连接环境与线粒体凋亡途径的纽带（Green，2011）。可见，线粒体途径的细胞凋亡是由环境信号所触发的。

细胞所处的环境往往偏离理想条件，细胞面临着 DNA 损伤、缺乏生长因子等各种物理或化学压力。这些压力通常会触发线粒体凋亡途径。当这些压力超过某些临界值时，细胞不能再保持其完整性，细胞就会通过凋亡而死亡。很明显，这种凋亡途径是因为细胞所受压力而被激活的。因此，细胞凋亡是生物体为了整个生物体更好地生存而迫使有害细胞死亡的过程。换句话说，凋亡不是细胞本身的自愿行为。因此，细胞凋亡与多细胞生物体的生存公理并不矛盾。

除了细胞压力外，脊椎动物和无脊椎动物的发育信号也会导致细胞死亡。在脊椎动物的肢体中产生手指足趾提供了一个生动的例子。鸡和鸭肢体发育的差别表明了细胞凋亡结果的特征性差别。鸭子的蹼脚可以帮助它们游泳，而鸡不会游泳，蹼脚在鸡的生活中不起任何作用。这种通过细胞凋亡而导致的细胞死亡机制是由为了生物体的需要而设计的遗传程序精确地控制着。在这种情况下，细胞也是被迫根据代表整个生物体有更好生存机会的遗传程序的要求而死亡。

另一种常见的细胞凋亡途径是死亡受体途径。在这种途径中，细胞外配体在一些被称为"死亡受体"的特殊分子上与细胞表面结合。死亡配体属于肿瘤坏死因子（TNF）家族，死亡受体属于 TNF 受体（TNFR）家族（Green，2011）。当死亡受体被结合时，它们将激活 caspase-8，从而引起 MOMP，并驱动线粒体凋亡途径。死亡受体途径是通过外部信号触发细胞参与死亡途径的一个例子。

类似地，炎症途径是由感染引起的信号所诱导。在具有免疫系统的生物体中，淋巴细胞和吞噬细胞可以检测到感染的细胞，并指示其在炎性途径中死亡。

当细胞发生突变而导致其失去群体抑制而增生成癌症时也会引发细胞死亡机制。癌细胞的存在对整个生物体的存活是有害的，因此有一些信号触发癌细胞死亡。例如，肿瘤抑制机制指定癌细胞死亡，这与发育信号的情形相似（Green，2011）。然而，癌细胞会经过代谢重新编程以维持生存和快速增殖。换句话说，癌细胞和其他细胞之间的区别是癌细胞能够建立足够强大的抵抗力来抵御来自更高层次组织的死亡命令。

根据生存公理，我提出一种细胞凋亡在细胞水平上的可能机制。当细胞受到诸如物理或化学压力或 DNA 损伤等环境的干预时会激活应变。细胞会尽力抵抗这些压力，并修复内部发生的损伤。然而，压力可能无法完全抵抗或损伤不能完全修复，那么，这些残余效应和损伤会积累并会使细胞变形。当这些效应和损伤达到一定阈值时，细胞的变形就会被生物体的防护系统所检测。这时生物体会发出一些信号来触发死亡过程，或者发送一些死亡配体与死亡受体结合，这样细胞凋亡便会发生。

2.4 生物学的目的性

目的性是科学哲学中的一个重要概念。目的性的研究早在亚里士多德时代就开始了。亚里士多德仔细观察了生物世界中的现象，并得出结论，所有这些过程都是由目的导向的。亚里士多德在他的目的因（Final Cause）中将这一结论推广到非生物世界，宣称生物世界和非生物世界中的所有物质都有一个潜在的目的（Pross，2012）。在现代科学革命中，我们对非生物世界的理解和知识得到了极大的提高，自然界的潜在目的性的观念被完全否定，取而代之的是自然界是完全客观的观念。显然，这种观念的改变主要是基于在非生物世界，更确切地说，是在物理世界中获得的科学知识。从逻辑上讲，现在所使用的方法在原则上与亚里士多德的方法是一样的，即把在一个领域中得到的结论不加验证就外推到另一个领域。现代科学革命的结果证明亚里士多德的外推是错误的。那么现在的外推是否正确呢？让我们回过头来深入探讨一下科学哲学。

自从科学革命以来，物理学取得了巨大的成就，但生物学研究发展缓慢，因此，科学哲学的概念和理论基本上是建立在物理学的基础上，属于物理学主义。相比之下，生物学对这些概念或理论的影响很小。物理学主义者本质上是还原论者，认为一切都是机械的和决定论的（Mayr，2004）。他们试图用物理科学的概念和原理来解释所有的生物过程。他们还试图将复杂的生物系统简化为简单的物理化学系统，并将生物学的概念和理论简化为物理学的概念和理论。对于那些不能简化为物理学的概念和过程，物理学主义者直接地将它们贴上不科学或不存在的标签（Mayr，1997）。目的性显然就被包括在内。他们声称，由于非生物世界中不存在目的性，并且也不能将其归结为物理学科学，所以在生物世界中也不应该存在这种目的性。

近几十年来，随着生物学的迅速发展，物理科学和传统科学哲学中的概念和理论的局限性越来越明显。为了满足发展的需要，我们不应该把从实验观察和理论分析得到的新思想和新概念局限于传统的理论框架。相反，我们应该修改和扩展那些过时的理论，以符合生物世界中的新发现。毫无疑问，生物世界中的目的性是普遍存在的，是不可否认的。我们不应该因为它与经典科学哲学不一致，就拒绝承认这种无处不在的特性。相反，我们应该修改和扩展现有的科学范畴，以涵盖生物学理论中的这种独特的性质。迈尔用新哲学的观点分析了目的性及其在生物学发展中的作用，表明所有生物世界中的目的导向过程都是程序化过程（Mayr，2004）。在进化过程中使用目的导向和认为进化过程或进化趋势是有目的导向都是正确合理的。过程的目的导向是由程序所控制，而这些程序可能是进化过程中自然选择的产物。他用目的性（teleonomy）来表示进化程序所控制的目的导向（Mayr，2004）。因此，迈尔阐明了进化生物学中的目的性。

最近，阿迪·普洛斯（Addy Pross）更详细地描述了目的性，但他主要还是将讨论局限在进化生物学的范围。我们每天都观察到有关生命系统和非生命系统的现象并且目睹了两者的巨大差别。普洛斯对此有生动的描述："所有的生物都表现得好像有一个议程。每一种生物都忙于自己的生计——筑巢，收集食物，保护年青一代，当然还有繁衍后代。……我们凭着生命具有目的特性的直觉来理解生物世界的运作，当然包括所有人类活动。……相比之下，在非生命世界中，理解和预测是在完全不同的原则基础上实现的。那里没有目的性，只有公认的物理和化学定律。"（Pross，2012）显然，目的性是生命系统与非生命系统的根本区别。生命系统有一个"议程"，可以代表自己行事。普洛斯试图用物理科学的定律和原理来解释目的性。他提出了动态动力学稳定性（dynamic kinetic stability, DKS）这个重要概念，并将目的性的本质归结为生命系统的能量收集能力。这种能力能够使生命系统摆脱能量最稳定状态的热力学约束，并能使其DKS最大化。生命系统需要消耗能量来达到DKS的最大化。这种能量能够抵抗热力学平衡，使生命系统保持在远离平衡的状态（Pross，2012）。他以汽车为例。具有能量收集能力的复制实体就像一辆带有发动机的汽车——它可以上坡，而不是像一辆只能下坡的没有引擎的汽车。

当然，普洛斯已经向前迈进了一大步，帮助我们从物理学的观点来理解目的性的基本特征。然而，我认为这只是故事的一部分。目的性并不只是指带有发动机的汽车的"上坡"行为。即使是带有发动机的汽车，它仍然需要一个司机或一些指令才能到达目的地。换句话说，要完成议程，具有能量收集能力的复制实体是不够的。它还需要决策能力来知道复制什么、何时复制以及如何复制，这是完成议程的关键部分。那么，只有能量收集能力的复制实体在何处以及如何得到这些信息？显然，能量收集能力只是复制实体完成议程的必要条件而已。目的性的本质部分是去哪里以及如何代表自己的利益行动。换句话说，能量收集能力仅仅是实践目的性的必要条件，而决策能力是实践目的性的关键部分。显然，目的性是无法只用物理科学的定律和原理来圆满解释的。生物学与物理科学的关系本质上是不同层次结构的关系，生物学具有较高的组织结构层次。正如尼尔斯·波尔（Niels Bohr）在20世纪30年代所指出的那样，生命现象应该用不同于物理学和化学的原理来解释（Bohr, 1933）。事实上，目的性属于生命系统的生物学属性，正如在第一章中所讨论的那样，是不能完全用物理科学的原理和定律来解释的。

复杂性在生命系统的特性中起着重要的作用。正如迈尔指出的，"在这个介观宇宙中，没有任何非生命系统的复杂性能够与由大分子组成的生物系统和细胞的复杂性相比拟"（Mayr, 2004）。尽管生命系统和非生命系统都是由相同"死的"化学元素组成，但生命系统的突现特性与非生命系统的突现性质有很大的不同。目的性就是这些突现特性之一，反映了生命系统的目的行为和活动（Mayr, 2004）。尽管我们还不了解"死的"分子是怎样结合变成具有目的性的活细胞，但是在生命世界里存在的目的性是不可否认的。事实上，目的性只不过是整体论的结果，也就是说，组分之间的相互作用而导致的突现特性，类似于水分子的水态特性来自于氢原子和氧原子的组合。因此，目的性一点都不神秘。有些人觉得神秘，也许只是因为它不能用物理科学来解释。

在某种程度上，目的性在解释生物学中的"终极因果关系"，即进化生物学领域中已经被接受了。生存公理指出，目的性也可以用来解释生物学中的"邻近因果关系"，即功能生物学。目的性不仅仅只是用来描述生命系统的行为或过程的简洁用语，它直接来自于我们对生命体系的观察和理解，是生命系统所特有的内禀属性。行为是否有目的是生命系统和非生命系统之间的根本区别。因此，生命系统的目的性应该被认为是生物学理论中最基本和最原始的元素，不能被还原到物理科学。

目的性在科学中尚未得到普遍接受的另一个原因是植根于物理科学的科学哲学的影响。传统的科学哲学认为，目的性是人的主观思想，因为科学只研究客观性，因此它不应属于科学研究的范畴。然而，生命系统的目的性与其行为密切相关，没有目的性这个概念，我们就无法理解它们。因此，生物学家面临着一个两难的境地。就像霍尔丹（Haldane）所描述的那样，"目的性就像是生物学家的情妇：他不能没有她，但他又不愿意和她一起出现在公众场合"。

我们该何去何从？首先，我们应该意识到传统科学概念对我们的思想和思维方式的局限。然后，我们寻找可能的方法来突破这些障碍，建立一个新的理论来描述生物世界的原理和规律。现代科学的发展表明，自然界的主观性和客观性并非如我们以前所预期的那样决然分开。以量子力学中的测量为例。一般情况下，量子系统的状态是一些本征态的叠加。对系统的测量（或观测）会导致系统的波函数坍塌为其中一个本征态。换句话说，如果我们想知道量子系统处于什么状态，我们就必须测量这个系统。在测量之前，我们只知道系统处于每个本征态的概率，但我们不知道系统确实会处于哪个状态。测量包括被测物体与测量仪器之间的物理相互作用，以及仪器与观察者之间的心理物理相互作用（Jammer，1974）。在这种情况下，客观状态取决于主观测量。著名的薛定谔猫悖论生动地说明了这一点。悖论说，一只猫被放在一个黑匣子里，在这个黑盒子里，一个专门设计的装置有50%的概率杀死猫。当我们打开盒子的时候，猫死或活的概率都是50%。然而，在我们打开盒子之前，根据量子力学的基本原理，猫处于一个不确定的状态，它可能是死了，也可能还活着，概率都是50%；也就是说，猫处于既死又活的状态。（Gribbin，1984）这并不是说，由于我们没有打开盒子，所以我们不知道猫的命运；而是说，由于我们没有打开盒子，猫本身处于一个生死不确定的状态。换句话说，猫的客观命运取决于我们主观的观察行为，这与我们日常生活中的经验是相矛盾的。在量子世界中，主观观察与客观状态是密切相关的。

量子纠缠是微观世界中一个有趣的现象。不管相隔多远，来源相同的两个量子粒子都涉及纠缠关系，即一个粒子可以瞬间知道对另一个纠缠粒子的测量。实验结果表明，一旦建立了量子纠缠关系，这种关系将保持不变，粒子可以区分和识别另一个纠缠粒子而不受时间和空间限制。这种现象不能用经典物理科学的概念来解释。事实上，微观粒子的这种特性类似于人类的意识。因此，我们也许可以认为量子纠缠的存在是微观粒子意识存在的证据。我们都知道意识的存在，我们感到困惑的原因是因为我们不能用时间、空间、质量和能量这些概念来描述和解释它。进一步的量子测量分析表明，意识不能被进一步还原。现在，越来越多的人相信意识就像时间、空间、质量和能量一样，是自然界中物质的另一个基本的、独立的特征。正如我们不能用空间来解释时间一样，我们也不期待意识可以被诸如时间、空间、质量或能量等基本物理量来解释。

再以精神分裂症为例。精神分裂症是一种以社会行为异常为特征的精神疾病，是认知能力缺陷所致。以前，诊断大多是基于行为的观察和经历的报告，通常没有客观的测试可用。脑成像技术如功能核磁共振成像（fMRI）和PET的研究表明，这种疾病与大脑结构异常有关。这个例子清楚地表明，主观的人类行为与客观的大脑结构密切相关。同样，我期待，"主观"的目的性将会在未来的大脑结构研究中找到其客观性。

从上述例子可以看出，随着科学特别是生物科学的进步，传统的主观性和客观性的定义之间的界限会变得越来越模糊。我们可以预期，主观性和客观性将会越来越紧密联系、相互影响，不能被完全分开。也许在生命系统的属性二象性中，主观性和客观性可以统一起来，即传统的客观性表现在物理化学属性中而传统的主观性表现在生物属性中。我认为，如果我们想研究生命的本质，寻找生物世界所特有的基本规律，我们就不可避免地要将目的性纳入我们的生物学理论中。没有目的性，我们也许只是研究生命系统的物理化学属性。

通过把目的性引入到生物学理论，我们就可以预测生命系统的一些行为和活动。例如，当溶液中存在葡萄糖时，我们可以预测细菌游向葡萄糖的轨迹；我们也可以解释鲑鱼的产卵迁移。

人们试图用物理学的原理和概念来解释目的性和生存公理的特征是很容易理解的。然而，从方法论上讲，这种努力极有可能是徒劳且令人失望的，其情形就如同试图用牛顿物理学的概念和原理去理解狭义相对论的光速不变原理一样。虽然光速不变原理和牛顿物理学的原理是不相容的，但它们毕竟还属于非生命世界。生命世界中的原理与无生命世界中的原理之间的差别预计要比它们大得多，因此，如果生命世界中的原理不能被无生命世界中的原理解释的话，那就不足为奇了。例如，意识难题，即经验和感觉，不能用物理科学来解释，那是因为它超出了物理科学的范围。罗森伯格（Rosenberg）认为，在实际中，目的论的解释不可能完全还原为非目的论的解释。一般来说，我们并不指望生命系统的生物学属性能够被物理科学的原理所解释。

2.5 生存公理的意义

一般来说，生存有两个含义，一个是物种的生存，与遗传信息从一代传递到下一代密切相关，属于进化生物学范畴；另一个是个体的生存，这与个体的行为有关，属于功能生物学范畴。个体的生存是物种生存的基础，而物种的生存是个体生存的保证。这两者互惠互利、密切相关。物种的生存被认为是遵循适者生存的自然选择。生存公理指出，生命系统的行为和活动是以在现有的环境中求生存为最终目的。

行为和活动的目的导向性普遍存在于生物世界中，从亚里士多德时代就开始被研究。目的论（teleology）被用来描述宇宙世界的目的导向性，包括非生物世界和生物世界，而目的性（teleonomy）被用来解释进化生物学中的目的导向性。这些研究大多局限于哲学讨论。在这里，目的性被用来解释功能生物学中的目的导向性，并用来描述生命系统具体和明确的行为和活动。

通过生存公理把目的性引入功能生物学的意义是什么？首先，引入目的性揭示了生命系统和非生命系统的根本区别。这里的目的性不仅仅是用来描述观察现象的一种简洁词汇，它是生命系统具有的内禀特性，也是生物学自主性的具体体现。牛顿运动第一定律将惯性引入物理学，这是物体在所有运动过程中的内禀特性。同样，生存公理把目的性引入功能生物学，这是生命系统在所有生物过程和活动中的内禀属性。其次，目的性的引入表明生命系统遵循与非生命系统完全不同的规律。我们有必要建立一个全新的生物理论体系框架。最后，寻求生存是生物世界的基本原则。没有求生本能，生命系统的行为和活动将无理可循。因此，生存公理是生物学理论的基础。没有生存公理就没有任何生物学理论可言。如果我们把生存公理应用到人类社会这样一个特殊的生物世界，上面有关生存公理的含义就很容易理解了。如果人类不是以生存为其最终目的，那么人类所有的法律和规章制度都将失去基础和意义。

2.6 生命的定义

生物学是关于生命现象的学科，旨在揭示生命的基本原理。因此，生命是什么是生物学的根本问题。自古以来，这个问题一直吸引着人类去回答。到目前为止，我们仍然没有得到明确和令人满意的答案。在过去的200年里，生命的定义很多，主要是基于与生命有关的生物功能，如遗传、生殖和新陈代谢。马尔切洛·巴比里（Marcello Barbieri）在书中列出了许多生命的定义（Barbieri，2003），在这里我只列举数例：

"我认为，这三个属性：可变性、自我复制和异相催化，包括了对生命定义的必要和充分条件。"（Norman Horowitz）

"生命是目的性机器、自我构建机器和自我复制的机器。换句话说，所有生物都具有三个基本特征：目的性、自主形态和不变繁殖。"（Mond & Jacques）

"生物系统是一种自我复制、自我调节和以环境能量为补给的开放系统。"（Sattler，R.）

"生命是一种化学系统，它能够通过自我催化而进行自我复制，并制造错误，从而逐渐提高自催化的效率。"（Brack & Andre）

还有一些是基于结构的定义。例如，"生命是当生物前化学系统的分子多样性超过复杂性阈值时所产生的一种突生现象"（Kauffman，1995）。斯图尔特·考夫曼（Stuart Kauffman）进一步解释说："在这个观点中，生命以整体出现，并且始终保持为整体。在这个观点中，生命并不是定位于它的各个部分中，而是存在于它们所创造的整体的集体新质中。……集合的系统是有生命的，而其部分只是化学物质"（Kauffman，1995）。

这些关于生命的定义也许还是未能令人满意。例如，活鱼和刚刚死的鱼之间的结构复杂性没有差异，但它们的行为表现出不同的特征，属于不同种类的系统。

为了弄清楚生命是什么，我认为我们需要知道生命系统和非生命系统之间的本质区别。从前几节的讨论中，我们已经知道两者的区别在于目的性。在这里，我提出一个关于生命的定义：生命就是有从事目的性行为的能力。

上述定义也许并不能完全涵盖生命的各个方面，但我认为，至少它抓住了生命的本质。而且，通过直接观察，可以很容易地识别一个系统是否属于生命系统。以病毒为例。病毒自身不能复制代谢,需要宿主细胞存活。但实验结果表明,病毒利用许多手段侵入细胞并在那里生存。显然,它的行为具有目的性。根据上述生命的定义,病毒属于生命系统。然而,按照生命的一般定义,"生命系统应该具有自我复制和代谢的能力",我们难以确定病毒是否应归入生命系统。

> 生命有序，死亡无序。
>
> ——马尔姆斯特伦（Bo C. Malmstrom）
>
> 衰老没有功能——它是对功能的颠覆。
>
> ——康福特（Alex Comfort）

第三章 细胞的有序性及其衰老

3.1 背景知识

我们都目睹生物体的衰老并且认为大部分（如果不是全部）生物体最终都会衰老。日常生活和研究实验中，对衰老的观察给我们的印象是：衰老是生物体本身自发的过程。这种情况使我想起了在非生命世界中对物理系统的类似观测：没有任何驱动力的物体会降低其速度并最终停止。这样的观察使亚里士多德得出如下的结论：必须对物体施加一定的力，才能使物体以匀速运动。这个结论被接受了大约 2000 年，直到伽利略根据他的实验结果指出，物体所受到的摩擦而不是物体的任何内禀属性导致物体减速和停止。牛顿把这一思想推广到他的运动第一定律中："每个物体都保持静止或匀速直线运动状态，除非施加力迫使它改变其状态。"换句话说，每一个物体都具备惯性用以抵抗任何试图改变其运动状态的尝试。那么生命系统中的衰老现象又如何呢？

在这一章中，我将研究生命系统有序性的本质和衰老的特性，最后讨论细胞的有序性与衰老的关系。

3.2 生命系统的有序性

生命系统具有高度的复杂性和精致的有序性。它具有结构自我组装、生理活性自我调节、增殖自我复制的能力。生命系统的这些特性归因于其复杂性和有序性。此外，生命系统还是一个开放系统，尽管它不断地与环境交换大量的物质和能量，仍然可以保持良好的动态平衡。这个特点的原因是相同的，那就是系统的高度复杂性和精致有序性。生命系统的另一个特点是它的层次等级结构。正是生命系统这些独特的特性使其独立于非生命系统,与非生命系统分开。例如,人体是由细胞、组织和器官组成的生命系统。它能生长、发育、增殖和新陈代谢。人体的层次等级结构是高度有序的,身体内的物理化学过程和生理活动具有高度合作性和精致协调性。人体生长时组织结构的有序性增强。因此,整个人体是一个远离热平衡的有序系统。如果人体的秩序被破坏,它的熵值便增加,那么生命就无法维持。

生命系统的有序性被普遍认同和接受,但对有序性的性质和来源却有不同的解释。如果把生命系统视为热力学系统,那么系统的有序性需要能量和"负熵"来维持;如果将生命系统看作钟表装置那样的机械系统,那么,分子的热运动就不足以破坏或改变系统的构象或结构(Schrödinger,1944),这样的话,生命系统的有序性并不需要能量或"负熵"来维持。至于有序性的性质和来源,薛定谔提出了两种不同的机制:(1)从统计学机制产生的"有序来自无序";(2)由新物理定律(或非物理定律或超物理定律)产生的"有序来自有序"(Schrödinger,1944)。

考虑到生命系统是一个具有有序结构的开放系统,它通常被认为是一个"耗散结构"。耗散结构是伊利亚·普里戈金(Ilya Prigogine)提出的,用来解释在热力学系统远离热平衡态的条件下,当其中一些参数达到阈值时由统计涨落引起的有序结构(Prigogine & Nicolis,1971)。这种有序性有时被称为"有序来自涨落"。尽管生命系统与热力学系统中的耗散结构有相似之处,如同样是与环境交换能量和物质的开放系统以及具有有序结构,但两者之间仍有一些本质的区别。只有当热力学系统远离热平衡态时,某些参数接近阈值从而热力学涨落能被放大时才会出现耗散结构。然而,生命系统如活细胞的有序性与生俱来,并不需要任何涨落或其他因素来诱发。有序性随着个体的生长而增加。而且,有序性使生命系统如细菌在细胞内有效合成和分配不同种类的蛋白质和其他分子,也能使哺乳动物的受精卵正确地发育为有机体。所有这些都可能是受到了细胞内基因组信息的指示。因此,我认为,生命系统的有序性通常具有层次等级结构,是经过漫长的自然选择过程而产生的,在本质上与统计涨落引起的热力学系统的耗散结构是不同的。生命系统的有序性应该是系统本身的内禀特征。

薛定谔将生命系统的有序性看成是钟表装置的有序性,并用来解释其"有序来自有序"的概念。他进一步说明了钟表装置和生物体之间的关系:

第三章 细胞的有序性及其衰老

"现在，我认为有必要多讲几句话来揭示时钟和生物体之间的相似之处，生物体也是依赖于一种固体——形成遗传物质的非周期性晶体——而大大地摆脱了热运动的无序。"（Schrödinger，1944）斯图亚特·考夫曼引入了"免费有序"（order for free）的概念，并提出"生物体中的许多有序可能根本不是选择的结果，而是自组织系统的自发秩序。有序，丰富而生机勃勃，不是为了对抗熵潮，而是无限量提供，支撑着所有后续的生物进化。生物体的有序是天然的，并不仅仅是自然选择的意外胜利"（Kauffman，1995）。他进一步讨论了细胞的基本特征，如细胞的体内平衡稳定性（比如说，生物惰性防止肝细胞转化为肌肉细胞），认为生命系统中的有序结构来源于自组织过程，如个体发育过程（Kauffman，1995）。

我赞同生命系统的"有序来自有序"的总体观点，因为它与传统的热力学平衡观念拉开了距离。但我认为，生命系统和机械系统是两种不同的系统，前者是动态的，后者是静态的。因此，这两种系统的有序性也应有所不同。首先，生命系统的有序性与机械设备的有序性的来源有着本质的不同。机械设备的有序性来源于工艺大师或钟表匠的设计，而生命系统的有序性则可以通过考夫曼提出的自组织过程来获得。其次，维持两种系统的有序性的机制应该有所不同。机械设备的有序性得以维持是因为热运动不足以破坏设备的结构。然而，生命系统是一个动态系统，内部有许多相互作用的成分。系统的有序性可能是由在进化过程中所获得的机制来维持。

生命系统的有序性是如何维持的？生命系统是一个开放系统，与环境不断地交换物质和能量。根据热力学原理，生命系统的有序性是由生命系统从环境中获得的能量和"负熵"来维持的。然而，据我所知，直到现在还没有直接的实验证据表明能量或"负熵"被用来维持生命系统的有序性。基于下面的讨论，我认为生命系统的有序性来自于系统本身的内禀属性，不需要外部的任何东西来维持，即"免费保持有序"。相反，生命系统的有序性需要外部环境的干预才能被改变。

确实，生命系统需要营养来维持它的生物活动。但是，从营养物或食物中获得的能量或"负熵"是用来维持生命系统的有序性吗？当食物是由生物大分子组成而废物是无机小分子时，人体无疑会得到一些"负熵"，因为一般来说，生物大分子比无机小分子更有序。然而，从维持人类生命的角度来看，氧气分子和水分子比食物更重要。在这种情况下，人体没有得到任何"负熵",因为氧分子和水分子并不比废物分子更加有序。

动物的冬眠可以提供一个令人信服的例子来说明外部环境和生物系统之间物质和能量交换的效果,帮助我们更好地理解生命系统有序性的维护。有些冬眠动物在冬眠季节可以不吃不喝,存活100天以上。换句话说,这些动物几乎没有从环境中获得能量或物质,但仍然可以通过极低的新陈代谢来维持它们的生命(推测是维持其组织和结构的有序性)。这一现象表明,当摄入外部物质和能量减少时,生命系统可以调节新陈代谢水平以保持组织的有序性。因此,生命系统的有序性不一定需要外部能量或负熵来维持,而能量很可能是用于生物化学反应和活动。这种情况可能类似于工厂的运作。工厂里的机器已经安装好了(当然,机器的安装需要能量),这些机器的运转和产品的制造需要能量和材料。当能量和材料减少时,机器将减慢工作进度,但机器在工厂的设置保持不变。上述讨论可以使我们得出结论:物质和能量的摄入是用于生命系统内部的新陈代谢,而不是用于维持有序性。因此,有序性是生命系统的内禀属性。

薛定谔把类似的讨论应用到在绝对零度下的钟摆(Schrödinger,1944)。钟摆周围的空气分子的热运动阻碍了钟摆的周期性运动,产生了热量,因此时钟需要能量来维持时钟的运转。然而,这种热运动不足以破坏时钟设备的有序结构。换言之,空气分子的热运动不能改变时钟设备的有序性。这种能量只用于时钟的运转。正如薛定谔所说,只有当设备被加热到融化时,钟摆的有序性才会改变。

热力学定律只适用于热力学系统。热力学系统以大量分子的随机热运动为特征。生命系统是热力学系统吗?以细胞为例。水分子和无机小分子如阳离子占据了细胞质内部热运动的大部分。然而,大多数生物大分子在细胞内构建各种有序结构,如细胞器,因此这些大分子并不参与随机热运动。此外,在生理条件下细胞内有序结构是相当稳定的,因此不会受到细胞内分子热运动的破坏。细胞内还有一些有序的运动如胞质环流,这种环流可以把细胞器、一些营养物质和代谢废物转移到细胞质中适当的位置。在正常生理条件下,细胞内的转录因子和酶等蛋白质自发地形成有序构象,细胞内的热运动不能将这种有序构象转化为无序构象,如无序卷曲构象。细胞内确实存在一些游离氨基酸和核苷酸,但这些生物单体多数参与高速有机合成,而不是参与太多的热运动。因此,细胞不同于通常的热力学系统。正如薛定谔所说,"生物体似乎是一个宏观的系统,其部分行为接近于当温度接近绝对零度、分子的无序度被消除时的纯机械（与热力学对照）行为，而所有系统都倾向于产生这种纯机械行为"（Schrödinger，1944）。如果细胞必须被视为是热力学系统，那么细胞内的细胞器等有序结构应该作为不涉及热运动的热力学系统的边界。上面的讨论表明，即使细胞内分子的热运动可能增加系统的熵，但也不会改变细胞内基本有序结构的主要特征。

应该指出，地球上的生命系统是一个开放系统，通常是在恒温和恒压的条件下。根据热力学理论，这种情况下最稳定的状态是吉布斯函数（Gibbs function）值最小的状态，而不是熵值最大的状态。更具体地说，吉布斯函数 G 被定义为：

$$G = U + pV - TS \tag{3-1}$$

其中 U、p、V、T 和 S 分别是系统的内能、压强、体积、绝对温度和熵。

通常，当 S 增加时，U 也会增加。G 值减小的净效应取决于 T，当 T 大时，熵 S 的增加会使得 G 值减小，而当 T 小时，熵 S 的增加可能会使得 G 值增加。也就是说，在恒温恒压下。生命系统的最稳定状态不一定是熵最大的状态。

生命系统具有快速进行有序生化过程的特点，如蛋白质和 DNA 分子的合成。显然，这些有序合成增加了系统的有序性，从而降低了系统的熵。另一方面，被损坏的蛋白质和不需要的蛋白质的降解会增加游离氨基酸的数量，从而增加系统的熵。与一般的热力学系统相比，生物系统涉及无序运动和有序合成，对系统有序性的净结果取决于这两种不同过程的贡献。

3.3 细胞的有序性假设

细胞是一个复杂但精致有序的系统。虽然细胞是一个动态系统，其有序性难以定量定义，但细胞有序性的存在是不可否认的。细胞的有序性也许可以定性地定义为细胞内部组织的有序排列和它们相互作用的有序网络。例如，细胞质不是均匀分布的，它含有许多高度有序的结构，如线粒体、核糖体、溶酶体等细胞器。细胞骨架是由真核细胞中的微丝、微管和中间丝组成的复杂网络，用于连接纤维和微管，实现细胞器的精确定位和迁移。细胞作为一个整体，具有许多有序的结构成分，它协调着细胞的生长、增殖、分化等生理活动。在这些过程中，细胞内有许多协调一致的生化反应，这些反应发生在正确的时间以配合生物功能的需要。因此，有序性在细胞的生物过程中起着至关重要的作用。下面的假设说明细胞在有序性方面的内禀属性。

有序性假设：细胞具有使系统的有序性最大化的趋势。

这个假设指出，细胞的过程是有方向性的，就是使细胞的有序性增加。当细胞在给定的条件下达到其有序性最大的状态时，细胞将保持这种状态，除非受到环境的干扰迫使其改变。由于环境的压力或其他干扰，细胞可能会降低其有序性。假设表明，当压力或干扰被撤除或减少时，细胞有增加系统有序性的倾向，并试图恢复到原有的有序状态。应该强调的是，这一假设并不意味着细胞总能达到系统可能达到的最大有序性。这个假设只是说细胞可以达到现有条件允许的最大有序性。因此，细胞所能达到的最大有序性随着细胞环境的不同而变化。

这一假设显示了生命系统和非生命系统之间的另一个根本区别。对于非生命系统，它具有达到最稳定状态的趋势，例如能量最小的状态（如动力学系统）或最大熵的状态（如热力学系统）。相对而言，对于生命系统，它具有达到有序性最大、功能最佳的状态的趋势（这里我们假定有序性越大，功能越好）。换句话说，具有最大有序性的状态是生命系统最有利的状态，它并不需要能量或"负熵"的输入来维持。这一假设在某种程度上得到了薛定谔的支持："生命似乎是物质的有序和有规律的行为，它并不完全基于从有序走向无序的趋势，而是部分基于得到保存下来的现有秩序。"（Schrödinger，1944）

这个假设有两个含义。第一个含义是细胞的过程是沿着自发地增加系统有序性的方向。从哲学的角度来考虑，这个假设类似于零力进化定律（the zero-force evolutionary law），即进化系统具有增加它的多样性和复杂性的趋势（McShea & Brandon，2010）。第二个含义是细胞内部有驱动力去抵抗来自环境的任何干扰。在这方面，这个假设类似于牛顿运动第一定律，即物体都有保持其运动状态的能力，这由物体的惯性来描述。与此相似，每个细胞都具有通过相应的生物化学反应来抵抗来自外部环境干扰的能力。这个假设表明，生命系统具有与非生命系统相似的能力，以抵抗任何试图改变系统状态的干扰。

上述讨论引出了以下有关有序性假设的推论：

推论 1：细胞具有抵抗环境干扰的能力。

来自环境的干扰，如物理或化学压力，可能会降低细胞的有序性。当细胞的有序性降低时，细胞的功能可能相应地减弱。在这种情况下，推论指出细胞有抵抗这些干扰的倾向，试图去纠正秩序上的偏差，恢复原来的有序性和功能。

细胞抵抗干扰的内禀特性可以用一个叫作"生存能力"的新概念来描述。生命系统的生存能力可以定义为它抵抗来自环境干扰的能力。细胞抵抗干扰的能力越强，细胞的生存能力越强。显然，生存能力应由其基因型和表型决定，并应反映在其生物过程的行为和生命周期中。例如，同一生物体中不同种类的细胞可能具有不同的生存能力，不同物种的细胞也可能具有不同的生存能力。

根据现有的细胞理论，所有细胞都来自现有细胞的分裂。因此，细胞的有序性可视为由生物进化而来的内禀属性。从本质上说，所有的细胞都有自己的有序度。对于给定的细胞，有序度与其功能密切相关。似乎合理的假设是，细胞的有序度越高，细胞的功能就越好。换句话说，当细胞的有序度达到最大值时，细胞就能发挥最佳的功能。可以预期，细胞分裂后应该有一个过渡期。为了改善其功能，新分裂的细胞需要调整内部结构，以增加其有序性，并从其初生状态转变为具有较高有序性的成熟状态。

3.4 细胞的衰老

150多年来，关于衰老的争论一直很激烈。为了有效治疗与年龄有关的疾病和延长健康寿命，我们需要对衰老的起源和机制进行研究。尽管关于衰老的起源和机理已有很多理论，但这些理论的结论仍存在较大的争议。

衰老是生物界普遍存在的现象，是生物体在生命过程中经过一段时间后发生的，伴随着生理活动功能和应激反应能力的下降。它通常是指随着年龄的增长，生物结构和成分发生退化的不可逆过程。细胞的衰老是指细胞停止分裂，并伴有结构异常和生理功能障碍的过程。衰老细胞主要表现为适应环境变化的能力和维持体内平衡的能力都下降。结构变化通常包括细胞体积减少、细胞内水分子减少、原生质体硬化、细胞收缩和失去正常形态。在原生质变化过程中，细胞核内出现固缩，其结构不清，细胞核与细胞质的比值减少，细胞核甚至会消失（Campisi, et al., 2013）。

关于细胞衰老的机制有许多理论，可分为两个不同学派：错误理论学派和遗传/程序理论学派。错误理论学派认为，衰老是由细胞内各种错误引起的，其中包括大分子交联理论、自由基理论、体细胞突变和DNA修复理论以及废物积累理论。大分子交联理论认为过量的大分子交联是导致衰老的主要原因。DNA交联和胶原交联会破坏细胞的功能，不饱和脂肪酸的氧化诱导脂蛋白之间的交联从而降低膜的流动性。自由基理论表明具有强烈化学活性的自由基能攻击生物体内的DNA、蛋白质和脂类等生物大分子，导致DNA断裂、碱基的交联和羟基化以及蛋白质变性等损害。体细胞突变和DNA修复理论阐明诱导和自发突变的积累引起功能基因的丢失，从而减少功能蛋白的合成，这种情况导致细胞衰老和死亡。例如，辐射引起的年轻哺乳类动物的衰老症状与正常老年个体的衰老症状相似。

遗传/程序学派相信，衰老是由基因决定的自然进化过程，其中包括复制衰老理论、程序衰老理论和长寿基因理论。在上述理论中，复制衰老理论是阐明衰老机制最为流行的理论。这一理论基于DNA复制过程中端粒的缩短。端粒缩短的机制很简单，DNA聚合酶需要一个RNA引物来启动其合成，而引物所留下的位置不能被DNA填充，因此每次DNA复制后端粒会变短。尽管端粒缩短在体外细胞衰老中起着重要的作用，但端粒缩短对体内生物体衰老的影响可能不如我们想象的那么简单，因为端粒酶在体内的活性和作用机制与体外有很大的不同。例如，单细胞真核生物，如纤毛虫（Greider & Blackburn, 1989; Collins, et al., 1995; Lingner & Cech, 1996）和酵母（Counter, et al., 1997; Nakamura, et al., 1997）在体内都有高水平的端粒酶表达，可以弥补在DNA复制过程中端粒的缩短。即使在成年期，虹鳟鱼和龙虾在所有被检查的器官中都有很高的端粒酶活性（Klapper, et al., 1998a, b）。

海弗利克（Hayflick）是第一个描述培养中的胎儿成纤维细胞的有限复制能力的人，并将这些发现解释为体外细胞衰老。有人进一步声称，体内细胞的衰老和体外培养的细胞停止生长是相关的。从这一解释得出结论：对多细胞生物体而言，细胞分裂的次数是个像时钟般的限制因素。当端粒长度缩短到一定阈值时，细胞就不能再分裂了。

尽管关于复制衰老的研究非常多，但体外研究与体内衰老的相关性一直存在争议（Cristofalo, et al., 2004）。例如，复制衰老与原位细胞衰老的直接关系的主要支持之一是假定培养的皮肤成纤维细胞（和其他细胞类型）的复制寿命随着供体年龄的变化而下降。然而，使用不同年龄健康供体的研究表明，供体年龄与培养的人类成纤维细胞的复制寿命之间存在很大的不确定性，且没有统计学意义（Cristofalo, et al., 1998）。另一个令人困惑的问题与"端粒老化假说"有关，该假说认为复制衰老可能受端粒缩短的调控。实验结果表明，在从单个个体建立的多个克隆中，端粒长度的差异很大（Allsop & Harley, 1995）。因此，上述数据并不支持供体年龄与体外平均端粒长度有直接关系。

葛森（Gershon）和葛森（Gershon）列举了许多观测，说明没有证据显示体内细胞的增殖能力存在"海弗利克极限"；相反，大量的证据指出在体内并没有增殖极限。因此，所谓的"增殖衰老"不能也不应该被用作体内衰老的模型（Gershon & Gershon, 2001）。哈里森（Harrison）进行了一系列移植骨髓细胞实验，观察到超过小鼠寿命所要求的增殖能力，而且，在大多数实验中，当细胞被移植到年轻个体时，来自老年个体的细胞和来自年轻个体的细胞增殖得一样好（Harrison, 1985）。葛森和葛森提出，"海弗利克现象"是滥用体外细胞的结果并指出端粒衰老理论是不可靠的。没有这种滥用，体内系统的增殖能力远远超过了维持有机体整个生命周期中细胞增殖的需要（Gershon & Gershon, 2001）。

有趣的是，一些永生细胞系没有检测到端粒酶活性，但仍有很长的端粒。这个观察意味着可能存在一种替代机制来延长端粒（Biessmann & Mason, 2003; Bryan & Reddel, 1997; Dunham, et al., 2000）。因此，端粒缩短的机制在体内是复杂的，"海弗利克极限"也许不能成为检测衰老的普遍标准。

总之，上述讨论表明，端粒缩短并不是复制衰老的唯一计时机制。根据实验模型，还有其他一些在限制正常细胞增殖潜能方面起作用的机制（Cristofalo, et al., 2004）。

有两种衰老模型是从进化生物学的角度提出的（Goldsmith, 2014）。Medawar模型表明，遗传漂移和突变积累导致晚期有益基因的丢失或晚期有害基因的出现。Williams模型认为衰老来自于某些基因的多效性，这些基因在生命早期是有益的，而在晚年则是有害的。关于衰老的确切遗传机制以及在衰老过程中起重要作用的特定基因的一些重要问题仍然有待解答。

关于衰老的起源，有两种截然不同的理论。在程序化衰老理论中，衰老是有目的的遗传程序，因为寿命过长会产生进化上的缺陷，因此在一段时间后，衰老程序被开启，以加速损伤的积累和减少修复的能力。在非程序化老化衰老理论中，衰老是由活性氧等损伤的积累引起的。一旦累积超过阈值，衰老过程就会开始。这两种理论各自都有实验结果的支持（Goldsmith，2014；2015），而衰老的起源问题仍存在争议。

衰老研究的实验结果存在许多混乱。这些混乱主要来源于研究中所使用的条件，如体内条件和体外条件有很大的差异。例如，虽然动脉硬化、前列腺增生等病理条件下的生物学特征与体外细胞衰老相似，但在体外培养细胞中发现的大多数标记物在体内衰老研究中尚未得到证实。另外，体外培养的人的体细胞的染色体端粒在每次细胞分裂后都会缩短，从而导致细胞分裂次数受到"海弗利克极限"的限制。然而，一些物种的小鼠的染色体端粒在它们的生命中保持相当的长度，没有观察到明显的缩短。

为什么体外细胞和体内细胞有很大的不同？研究表明，造成这种差异的主要原因是体内细胞和体外细胞的环境差异。例如，在一般培养条件下，细胞在氧气浓度为20%的条件下孵育，远远高于生理条件（3%），此外，体外培养的细胞发生转移需要胰蛋白酶处理的细胞。相比之下，体内细胞可能不会遇到这种情况。胰蛋白酶处理不仅破坏细胞间的连接，而且还能消化细胞表面的受体。而且，最大的区别可能在于体内细胞处于三维生理环境中，而用于衰老研究的培养细胞往往仅限于二维。这使细胞在接收各种信号方面有很大的差异，无疑也会影响细胞的行为。迄今为止，还没有实验证据表明应激引起的体外复制衰老或过早衰老与体内组织或个体的衰老直接相关。利用体外细胞来研究体内衰老的机制确实是一个巨大的挑战。

3.5 有序性与衰老

根据普遍接受的定义，当细胞的功能降低到某些阈值以下时，细胞衰老的过程便开始。大多数细胞在衰老时其功能会改变（Campisi，2000）。结构和功能是生命系统的两个重要方面，它们密切相关。当细胞的结构被破坏时，细胞的功能就会相应地被破坏。显然，细胞的有序性同时降低。因此，细胞功能的下降是由于细胞有序性的降低。

细胞是一个开放系统，参与和环境的相互作用，并始终受到环境的干扰。物理和化学压力等干扰可能会扰乱细胞的有序性并降低系统的有序性。当这种破坏作用积累到一定程度时，细胞在生物过程中就会偏离其正常途径，其功能也随之下降。因此，细胞衰老是受到触发而开始。然而，每个细胞都有抵抗环境干扰的能力，而不同细胞能力不同。这种能力的差异导致不同细胞开始衰老的时间有先有后。

我们在日常生活中观察到，如果没有力的作用，任何运动的无生命物体都会减速，最终停下来。然而，牛顿运动第一定律指出，物体具有保持其运动状态的惯性，是摩擦力改变了它们的运动状态，使它们减速和停下来。同样，在生命世界中，任何生物都会经历衰老和死亡。这些无处不在的过程在生命世界中遵从什么因果关系？有没有什么驱动力可以阻止生物体的衰老和死亡？

一方面，生命系统的正常功能取决于系统的适当有序性。另一方面，生命系统的衰老又表现为结构和功能的退化。人们普遍认为，衰老始于细胞内的损伤累积超过某些阈值（Seregiev, et al., 2015）。这种现象可以用有序性假设来解释。正如第3.3节所讨论的那样，细胞的有序性包括系统中各组织的有序安排和连接所有组织的有序网络。环境的干扰和压力会对系统造成破坏，扰乱系统的有序性，从而干扰其生物过程。根据有序性假设，细胞具有增加其有序性的内禀属性。如果这些损害不是立刻致命的话，系统就有能力抵御系统中发生的任何进一步损害。在这种情况下，系统仍然可以保持在正常的生物途径，并发挥适当的生理功能。但是，当损伤的积累超过阈值时，这种破坏将超过系统的承受能力，系统的有序性会崩溃。因此，系统的生理过程将偏离正常的途径，衰老随之而来。这种情况类似于在无生命世界里移动静止的物体，如果施加在物体上的力不能克服静摩擦力，物体就不会改变其运动状态。然而，一旦驱动力超过最大静摩擦力，静止物体就会沿着驱动力的方向前进并加速运动。

上述讨论不仅解释了细胞衰老的机制，而且引出了以下有关有序性假设的推论。

推论2：当细胞的有序性降低到某些阈值以下时，细胞便开始衰老。

从这个推论可以看出，细胞有序性假设支持衰老错误理论，因为细胞具有不需要任何付出就能保持有序性的内禀特性。因此，细胞的衰老不是来自细胞内的遗传程序，而只能来自环境的干扰和损伤。

细胞可能遭受多种外部压力，因而导致自由基过多和代谢废物积累等细胞异常。这些异常的后果包括 DNA 断裂、蛋白质变性和核苷酸碱基配对特异性的丧失。所有这些异常都反映了细胞有序性的破坏。例如，蛋白质的变性可以被看作是将分子的有序构象破坏成无序构象。生物大分子的交联和碱基配对特异性的丧失可以被认为是对细胞网络有序性的破坏。细胞有序性的降低必然导致细胞功能的降低。因此，这个推论指出，细胞有序性的破坏是细胞衰老的根本原因。

实验结果表明，一般来说，如果细胞正常分裂，细胞衰老就不会发生。这个现象可以用有序性假设的推论来解释。在细胞分裂过程中，细胞内积累的损伤被分配到两个子细胞里，因此细胞分裂使得子细胞内的损伤积累的程度减少约一半。如果细胞继续分裂，细胞分裂的速度快于损伤积累的速度，则根据推论2，细胞就不会开始衰老。换句话说，如果环境干扰的影响不足以干扰细胞分裂，那么细胞就不会开始衰老。

每个细胞都梦想成为两个细胞。

——弗朗索瓦·雅各布（Francois Jacob）

第四章
细胞生长和细胞分裂原理

4.1 背景知识

为了生存，细胞必须生长和分裂。细胞生长和分裂是细胞生命的重要组成部分。事实上，它们是生命系统和非生命系统之间最根本的区别之一。细胞理论表明，目前所有的细胞都来自于现存细胞的分裂。细胞分裂主要有有丝分裂和减数分裂两种不同类型，它们承担着不同的任务，对生物体的生存起着不同的作用。一般来说，体细胞参与有丝分裂，生殖细胞参与减数分裂。不同类型的细胞分裂产生不同的结果。从理论上讲，有丝分裂后，两个子细胞具有相同的基因组，与母细胞的结构和功能基本相同。有丝分裂的结果是增殖。有丝分裂产生的子代细胞可以继续生长，从而使细胞的总体积增加（两个子细胞的体积与一个母细胞的体积相比）。相比之下，减数分裂不仅使子细胞染色体数目减少一半，而且在交换和自由组合过程中也改变了子细胞染色体的成分。减数分裂后的四个子细胞通常不会继续生长。减数分裂的结果是子代细胞遗传变异的多样性，而不是细胞总体积的增加。显然，参与有丝分裂和减数分裂的细胞具有不同的功能，表现出不同的特征，因此应遵循不同的原理。

就细胞大小而言，真核细胞通常大于原核细胞，大多数真核细胞的大小在 $10\mu m \sim 100\mu m$ 之间，而大多数原核细胞在 $1\mu m \sim 10\mu m$ 之间。细胞在多细胞生物体中的大小一般是在 $20\mu m \sim 30\mu m$ 之间。显然，植物和动物都不采取通过增加细胞体积来实现个体成长的策略。相反，它们采取通过连续细胞分裂来增加细胞数目的策略。体细胞体积表现出一个简单的周期性变化：有丝分裂后刚形成的子细胞在开始时只有母细胞一半的大小，但它们通过合成原生质能迅速地把体积增加到和母细胞一样大小，接着又开始另一轮细胞分裂（Xie, 2013）。

本章将讨论两种不同类型的细胞分裂，并对基本过程的机理进行研究。

4.2 细胞分裂与生存公理的关系

初看起来，生存公理和细胞分裂存在着一些矛盾。在细胞分裂过程中，一个亲代细胞消失，两个子代细胞形成。矛盾在于如何看待亲代细胞的命运以及亲代细胞与子代细胞之间的关系。换句话说，这种矛盾与生命的定义或细胞的寿命有关。在这里我提出一个解决方案。无论是有丝分裂还是减数分裂，一个亲代细胞分裂成两个子代细胞，并不是一个生命的结束和两个新生命的开始，而是一个生命的延续。每个分裂形成的子代细胞应被视为前一代亲代细胞的一部分。有丝分裂与减数分裂的区别在于，有丝分裂中的子代细胞与亲代细胞具有相同的遗传组成，而减数分裂的子代细胞与亲代细胞有不同的遗传组成。因此，如果其中一个子代细胞死亡，而另一个子代细胞还活着，则亲代细胞的生命仍在继续。更具体地说，如果细胞 A 分裂成两个细胞 B，而一个细胞 B 又分裂成两个细胞 C，等等，这样，我们有一系列的细胞，它们都起源于细胞 A，我们可以把所有这些细胞定义为"A 组细胞"。只要"A 组细胞"中有一个细胞还活着，那么，"A 组细胞"作为一个整体的生命仍然在继续。"A 组细胞"的寿命应从 A 细胞诞生之时算起直到"A 组细胞"中最后一个细胞死亡。这样定义的寿命可以特定地称为群体寿命。

对于像人类这样的多细胞生物体来说，群体寿命的生物学意义应该是指从受精卵的形成到生物体中最后细胞的死亡。对于细菌等单细胞生物体来说，即使每个分裂的细胞是一个独立的实体，我们仍然可以像在多细胞生物中那样，定义来自某些特定细胞的群体细胞的群体寿命。然而，在这种情况下，在选择起始细胞时存在着一些任意性。但对于这些不同的群体，确实存在着一些等级关系。尽管它们可以独立生活，但我推测，起源于同一细胞的每个细胞之间应该有着密切的联系，因为它们是整个群体的一部分，这一点得到细菌菌落、细菌生物膜（Nadell, et al., 2009）和细菌群感效应等现象的支持。这种同一起源的细胞具有紧密联系的现象可称为细胞纠缠，类似于神秘的量子纠缠现象。量子纠缠是指两个来自同一起源的微观粒子如此紧密地联系在一起，以至于当其中一个粒子受到某种相互作用时，另一个粒子就能即时感觉到。

根据上述讨论，可以提出细胞分裂与生存之间的关系。如果细胞不分裂，来自环境的破坏或压力可能累积，导致细胞衰老和死亡。细胞生存的命运仅由该细胞决定。然而，如果细胞不断分裂并形成一组群体细胞，那么即使群体中有许多细胞死亡，只要一部分细胞还活着，那么整个有机体仍能存活。在这种情况下，生存的命运是由群体内的所有细胞决定。显然，细胞分裂过程为生物体提供了更好的生存机会。因此，细胞分裂不是与生存公理矛盾，而是支持生存公理。

4.3 有丝分裂和细胞生长假设

为了生存和保持内部的稳定状态，细胞需要来自环境的营养物质等资源。通常细胞吸收的资源比维护生存和稳态所需要的多。事实上，细胞吸收了额外的资源来进行生长和分裂。生长和分裂是细胞的两个基本过程。细胞生长是一个高度调控的过程，通常伴随着细胞分裂；而细胞分裂通常被认为是控制细胞大小的手段。我们对这两个过程的理解已经取得了很大进展，对有关机制也有详尽的描述（Schmidt，2004）。然而，我们对这些过程的驱动力知之甚少。让我们在本节中详细研究细胞生长和有丝分裂。

首先，让我们考虑以下的思想实验。我们将一个活细胞和一个热力学系统分别放入相同的培养液中（如图4.1所示）。假定热力学系统与细胞有相同的细胞膜，包含相同的材料。这两个系统不同之处在于活细胞在细胞膜内具有有序结构，属于生命系统，而热力学系统不具有预先设定的有序结构，属于非生命系统。对于热力学系统，经过足够长的时间后，系统与培养液处于平衡状态，处于最稳定的状态，系统的大小不再随时间而变化。相比之下，活细胞在生长，其大小在增加。当细胞的大小达到某些值后，细胞就会分裂。分裂的细胞会再次生长和分裂。这一过程会一直持续到细胞数量达到饱和为止。如果取出一滴这样的饱和溶液（含饱和细胞的培养液）并放入另一个相同的培养液中，该细胞将再次生长和分裂。细胞的数量将呈指数增长，并将再次达到饱和。这个过程可以连续重复，直到达到细胞分裂的"海弗利克极限"（如果确实存在）为止。

上述思想实验的结果清楚地表明了生命系统与非生命系统的根本区别。尽管这两个系统的组成是相同的，但它们的构象和结构在其行为和特征上表现出巨大的差异。这一思想实验说明了细胞在有丝分裂过程中的内禀特性，并为细胞在有丝分裂过程中可能遵循的原理提供了一些线索，如

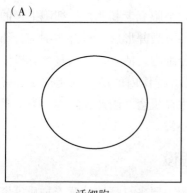
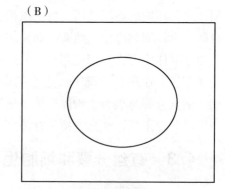

图 4.1 活细胞与热力学系统比较

注：在相同的培养液中活细胞和热力学系统的比较。(A) 活细胞；(B) 具有相同膜和含有与活细胞相同材料的热力学系统。

以下假设所述：

有丝分裂和细胞生长假设：体细胞具有将外部资源转化为内部资源最大化的趋势。

这里的资源定义为合成诸如核酸、蛋白质等生物大分子所需的营养物质和材料，以及其他一些小分子如金属离子。根据其相对于细胞膜的位置，资源可以分为外部资源和内部资源：如果位于膜外，则为外部资源；如果在膜内，则为内部资源。该假设表明，体细胞具有尽最大可能将外部资源自发转化为内部资源的内禀能力。为了这个目的，细胞从环境中吸收资源，从而增加其体积。当细胞生长时，体积增大，表面积与体积的比值减小。由于细胞与环境之间的交换效率取决于表面积与体积的比值，因此交换效率随细胞大小的增大而降低。为了使效率最大化，体积较大的细胞会分裂成两个体积较小的细胞，从而使表面积与体积的比值增大。换言之，细胞采用分裂的策略，最大限度地吸收外部资源。

这一假设指出了活细胞的另一个内禀特性。这种内禀属性是生物学自主性和"整体大于各部分之和"的整体论的表现。显然，这种内禀属性属于生命系统的生物学属性，因此不能用物理科学的原理或定律来解释。

应该指出，这一假设仅仅意味着细胞具有将外部资源最大限度地转化为内部资源的内禀趋势。然而，这一趋势未必会得到最大程度的实现，因为这一趋势的实现取决于环境的影响，而环境对这一趋势的实现起着重要的作用。例如，当空间不足或环境中资源不足时，接触抑制增殖将抑制细胞分裂。当细胞仅仅是组织或结构的组成部分时这种影响就更加明显，这时细胞的分裂能力很可能受到系统功能或结构要求的限制。例如，组成人体器官的细胞只能生长和分裂到一定程度，这应归因于人体的功能或结构的要求。对于单细胞生物来说，这一趋势通常在最大程度上能够实现。

你也许会问："体细胞为什么要将外部资源转化为内部资源？"我认为其原因是，内部资源可以由细胞自己利用和操纵，而外部资源则不能；内部资源是细胞的财产，而外部资源却不是。这一假设的意义在于转换的最大化为细胞提供了最大的生存机会。正如4.2节所讨论的那样，细胞分裂会产生一系列细胞而形成了"群体细胞"。群体内细胞越多，群体存活的机会就越大。因此，当细胞最大限度地将外部资源转化为内部资源的同时，也会使其生存的机会最大化。从这个意义上说，细胞是自私的，它为了更好的生存机会而寻求更多属于自己的资源。

细胞生长和细胞分裂是两个密切相关的过程。长期以来存在这样一个问题：是生长导致了分裂还是分裂导致了生长？根据这一假设，这两个过程共同合作，其共同目的是更有效地将外部资源转化为内部资源最大化。因此，这两个过程可视为一个过程的两个不同阶段。生长的目的是为了分裂，因为分裂会产生更多的细胞，从而得到更好的生存机会。分裂是为了更好地生长，因为分裂的细胞能更有效地吸收外部资源。可以推断，如果细胞不能分裂，细胞就不会持续生长，因为转换效率会随着细胞体积的增加而降低。

上述讨论引出了以下的推论：

推论3：分裂能力是细胞的内禀特性。

根据假设，细胞有将外部资源转化为内部资源最大化的趋势。为了实现这一趋势，细胞必须进行分裂。换句话说，细胞必须自发分裂，以便更有效地吸收外部资源。因此，细胞本身必须具有分裂能力。

终末分化的细胞通常不分裂，这似乎与推论相矛盾。在我看来，这一现象与推论并没有冲突，因为推论仅仅表示细胞的内禀能力。然而，如上所述，这种能力会受到环境影响的限制。事实上，这种现象只表明我们没有观察到终极分化细胞的分裂，这并不一定意味着这些细胞已经失去了分裂的能力。细胞没有分裂可能是由于环境造成的接触抑制增殖，或组织的结构或功能要求造成的其他限制，从而阻止细胞分裂。我们许多人在培养细胞时可能有以下的经验：当细胞浓度接近饱和时，细胞的数量没有明显增加，这意味着大部分细胞不分裂。然而，如果将一小部分培养物转移到另一个新鲜培养液中，我们发现浓度很快又会呈指数增长，说明细胞分裂能力已经恢复。同样地，当肝脏被手术切除其质量的70%时，剩余的组织会生长以恢复原始质量和功能，这表明即使当肝脏的大小正常时肝细胞没有分裂，但仍然保持着分裂的能力。这些结果意味着终末分化的细胞仍然有分裂的能力，当条件允许时，这种能力就会显现出来。

推论4：细胞分裂时的体积大小可以在一定范围内变化。

为了进一步讨论这个推论，我们可能需要引入一些新概念。细胞的组分可分类为必需组分和非必需组分。必需组分定义为细胞生存所需的必要成分。缺少任何这些必需组分，细胞就无法生存。有理由认为细胞的必需组分在细胞生长中具有较高的合成优先权。

细胞生长和分裂的目的是将外部资源转化为内部资源。然而，细胞在什么体积或细胞在多大时才分裂？从有丝分裂和细胞生长假设出发，细胞分裂时的体积越小，转化效率越高。然而，从生存公理来看，只有当子代细胞能够存活时，细胞才会分裂。换句话说，细胞只有在细胞所有的必需组分被合成后才能分裂。因此，这两个因素决定了细胞分裂的实际时间或细胞分裂时的体积大小。

让我们更详细地考查一下这个过程。实质上，细胞生长是核酸和蛋白质等细胞组分的复制过程，而细胞分裂则是这些组分的分配过程。细胞通过其表面与环境交换资源，其交换率由表面积与体积之比决定。随着细胞大小的增加，交换率会减小。在相同体积下，两个小细胞的表面积之和大于一个大细胞的表面积（在球体的情况下，$S_2/S_1 \sim 1.26$，其中 S_2 是两个小细胞的总表面积，S_1 是一个大细胞的表面积）。根据有丝分裂和细胞生长假说，细胞具有在体积小时分裂的倾向。

那么，为什么细胞只有在其体积或大小达到一定值后才会分裂呢？原因是细胞遵循生存公理。为了生存，细胞需要拥有细胞所有的必需组分，例如重要的生物大分子和其他一些重要的组分。如果一个细胞在复制所有的必需组分之前就分裂，则至少一个子代细胞必须死亡。基于生存公理，这一过程不应发生。细胞只有在其所有必需组分复制完之后才会分裂。有实验结果支持这个观点：当培养的变形虫将要分裂时，细胞质的一部分被切断。变形虫随后没有分裂，而是继续生长。如果一段细胞质再次被切断，则变形虫还是不会分裂。但是如果让它继续增长，当体积达到一定大小时，它就会分裂。因此，细胞只有在所有必需组分都被复制后才会分裂。换句话说，细胞分裂的必要条件是细胞的所有必需组分都被复制，而这也就是细胞可以进行分裂的最小体积。然而，这种情况可能不是细胞分裂的最佳条件，因为细胞中的某些辅助成分可能更有利于子代细胞的生长，这些可能取决于细胞和环境之间的相互作用。这些辅助组分的不同组合可构成一系列细胞分裂的最佳条件。因此，细胞分裂的体积或大小可以在一定范围内变化，其下限由生存公理决定，上限由细胞分裂的最优化条件决定。

4.4 减数分裂和细胞遗传多样性假设

有丝分裂和减数分裂是真核细胞分裂的两种主要不同形式。有丝分裂过程中，体细胞生长和分裂产生更多的细胞，所有细胞的基因组基本相同。分裂后的细胞仍保持分裂能力。减数分裂过程中，生殖细胞不仅改变了染色体的倍性，如人类的二倍体变成了单倍体，而且还改变了染色体中的构成。动物和许多植物的配子通常不参与有丝分裂。例如，人类的成熟精子和卵子不参与有丝分裂。

减数分裂是生物体在不断变化的环境中生存的重要机制。它产生了许多遗传变异，其中劣等的将根据达尔文提出的自然选择原理被淘汰。遗传变异来自第一次减数分裂的两个不同过程：交换和自由组合，前者先发生，后者后发生。在交换过程中，同源染色体间DNA片段的交换产生了不同于亲本染色体的新版染色体。在自由组合过程中，同源染色体随机出现在赤道板，并在纺锤丝牵引下随机向细胞两极迁移。对于人类来说，每一个细胞中23对染色体的自由组合可以产生2^{23}或超过800万套染色体组。尽管这一过程并没有产生新的基因版本，但它为子代提供了数量巨大的新染色体组合。为了提供更多的遗传变异，性染色体X和Y在减数分裂过程中起着类似于短的同源染色体的作用。上述两个过程的总效果是导致了大量不同的遗传变异产生，为真核生物群体适应环境的变化提供了遗传多样性。减数分裂后的配子染色体在基因上是各不相同的，有来自双亲的各种DNA组合。

在大多数生物体中，生殖细胞来自受精卵的第一次不对称分裂。因此，胚胎发育过程中的生殖分化是在胚胎发育的早期阶段决定的。种质成分不均匀地分布在未受精的卵细胞后极处。随着受精卵的第一次不对称分裂，含有种质的子代细胞将分化为生殖细胞，也称为原始生殖细胞。生殖细胞通过有性生殖产生后代，并将遗传信息传递给后代。在配子发生过程中，原始生殖细胞通过性别决定和分化产生配子。在这个过程中，细胞经历减数分裂，染色体配对，重组并改变子代细胞的基因型。精原细胞和卵原细胞通过减数分裂分别产生成熟的精子和卵子。在精子形成过程中，经过两次减数分裂后，一个精原细胞产生四个精子。虽然这四个精子形态相似，但基因型不同。由于精子不能通过有丝分裂增殖，而且所有精子都只能来自精原细胞的减数分裂，所以它们的基因型都是各不相同的。同样，成熟的卵细胞和三个极体是从卵母细胞经过两次减数分裂产生的，所有的卵细胞都有不同的基因型。这种产生生殖细胞的模式确保了配子基因型的多样性。

上述讨论引导出以下关于减数分裂机制的假设：

细胞遗传多样性假设：生殖细胞具有将其遗传变异多样性最大化的趋势。

这个假设表明，生殖细胞的主要功能是为后代提供遗传变异，而不是从环境中吸收资源。为了使物种在不断变化的环境中有更好的生存机会，生殖细胞具有把其遗传构成的组合最大化的趋势，这些组合可以在受精过程中被选择。因此，减数分裂为物种在不断变化的环境中生存提供了一种机制。因此，从哲学的角度来看，生殖细胞的这一特征与达尔文建立的进化论中自然选择的要求是一致的。

高质量精子的选择是在受精过程中进行的。以哺乳动物为例。哺乳动物在一次射精中的精子数量高达上亿。这些精子必须游过阴道、子宫，穿过宫颈进入输卵管，然后才可能与卵细胞接触。此外它们还需要穿过卵丘细胞和卵子周围的放射冠，并进一步通过称为透明带的结构使卵子受精。只有少数精子能穿过放射冠，通常只有一个精子能穿过透明带并使卵子受精。在受精的征途中，数以万计的精子必须与对手竞争赛跑，克服许多极端情况，如雌性生殖道中强酸条件。大部分精子在这个过程中死亡，并被溶解和吸收在雌性体内。这种在母体内选择精子的机制确保没有任何劣质精子能使卵子受精。显然，受精过程中精子的选择对生产优质后代起着至关重要的作用。这确实是达尔文提出的在进化过程中适者生存之自然选择的生动缩影。

由于精子和卵子在受精过程中的作用不同，它们在减数分裂过程中的行为方式差别很大。青春期后，精子每天产生数亿，而卵子则滞留在减数分裂Ⅱ中期，只有通过受精才能继续减数分裂。为了生产出高质量的后代，必须满足两方面的要求：(1) 具有尽可能多的遗传变异的精子可供选择；(2) 卵细胞本身具有高质量。为了提供尽可能多的遗传变异给卵子选择，正如上面所讨论的那样，精子的产生方式使得所有的精子都有不同的基因型。换句话说，为了保证遗传多样性的最大化，成熟的精子不参与有丝分裂，因为有丝分裂产生出具有与亲代细胞相同基因型的子代细胞。由于卵细胞不具备与精子相似的选择过程以防止劣质卵细胞受精，因此，它们的产生与精子截然不同。卵母细胞经历不对称减数分裂，只产生出一个卵母细胞和三个小而寿命短的极体。这个卵子从卵母细胞中获取所有的资源（细胞质，线粒体）。高质量的卵细胞以这种方式产生。

综上所述，为了生产高质量的后代以便能在不断变化的环境中生存，生殖细胞具有使遗传变异最大化的内禀属性，而选择机制则确保劣质配子不会参与受精。

4.5 细胞分裂的经济原则

正如前面所讨论的那样，活细胞在任何可能的情况下都有分裂的倾向。然而，在某些情况下，细胞并不总是分裂，如神经细胞和卵细胞。这种现象可能是由于细胞分裂的下列原则所致：

细胞分裂的经济原则：如果分裂的细胞不履行任何功能或对生物过程没有任何贡献，那么细胞就不会分裂。

经济原则表明，细胞只有在必要时才会分裂。尽管细胞具有内禀的分裂能力，即在可能的情况下进行分裂，但细胞分裂受到其环境和组织的限制，特别是当它只是组织的一部分时。换句话说，细胞只有在其分裂的子代细胞被环境允许或有益于生物体的功能和活动时才会分裂。这个原则是对细胞内禀分裂能力的调节机制，可视为细胞分裂过程的负反馈。对于单细胞生物体的细胞来说，由于每个细胞都是一个独立的实体，其分裂在许多情况下只受到环境的影响，因此这个原则对细胞分裂的限制通常不大。相比之下，多细胞生物体的细胞分裂受到细胞所具有的功能的强烈影响。例如，在多细胞生物体中，每种细胞都有自己的任务，通常以组织或器官的形式来完成。每个组织或器官都有其最佳尺寸来执行功能。当细胞数目达到一定值时，更多的细胞不能增强组织或器官的功能，这时细胞就不会继续分裂。这一现象是由细胞分裂的经济原则所造成的。

接触抑制增殖（CIP）是细胞相互接触时细胞生长停止的现象。最近的研究表明，这一现象是由细胞与细胞黏附受体启动的（McClatchey & Yap，2012）。在培养细菌的过程中，细胞的数量在开始时呈指数增长，然后尽管营养仍然充足，一旦细胞数量达到饱和就不会再分裂。单细胞生物的 CIP 可能是由于空间限制，使得进一步分裂的细胞不能成功存活。根据生存公理，细胞在这种情况下是不会分裂的。相对而言，在多细胞生物的组织形成或器官形成过程中，细胞在组织或器官正常形成时停止分裂，这可能是细胞分裂的经济原则所造成的结果。

细胞分裂的经济原则清楚地表明，尽管细胞具有内禀的分裂能力，但这种能力受到严格的调控，有时还会受到环境的严重抑制，特别是当细胞只是组织的一部分时。细胞在不同的组织或器官中发挥着不同的功能，因此在细胞分裂中表现出不同的行为。例如，如果细胞衰老得早，并且在生命中容易受损，如血细胞，那么它们就会分裂得快而且频率高。在这种情况下，细胞分裂以取代那些在组织中不能正常工作的细胞。另一方面，细胞分裂在细胞持续时间较长且不易受损的组织中就不易被观察到，如中枢神经系统中的某些神经细胞。因此，根据经济原则，这些神经细胞不需要增殖。卵细胞在受精前不继续第二次减数分裂的原因是因为卵细胞的功能是产生后代，但没有受精而继续第二次减数分裂显然不能履行这一功能。根据经济原则，卵细胞不会继续减数分裂。这些现象强调细胞只有在必要时才会分裂。

细胞分裂的经济原则也可以解释干细胞在具有等级组织的生物体中的某些行为。研究表明，几乎所有器官都有相应的干细胞，如肝干细胞、神经元干细胞。这些干细胞通常在正常的生理条件下保持休眠状态。根据经济原则可以推测，当终末分化细胞受损并需要补充时，只要同一类型的终末分化细胞能够增殖，干细胞会保持休眠状态。只有当这些终末分化的细胞不能完成这项任务时，干细胞才会参与有丝分裂并增殖。癌症干细胞也是如此。癌症干细胞通常处于休眠状态，只有在大多数癌细胞被化疗杀死时才参与有丝分裂和分化。这可能是有效治疗癌症的困难所在。

> 我不能创造的事物，我就不能理解。
>
> ——理查德·费曼（Richard P. Feynman）

第五章 细胞分化原理

5.1 背景知识

多细胞生物体通常由许多不同类型的细胞组成。同一生物体内不同类型的细胞在结构和蛋白质组成上都是不同的。它们在生物过程中发挥着不同的功能。然而，大多数这些不同类型的细胞都具有基本相同的基因组。它们通常是由同一个细胞如受精卵经过细胞分裂和分化而产生的。细胞分化是细胞从一种细胞类型转变为另一种细胞类型的过程。这一过程极大地改变了细胞的特征，如细胞的大小、形状和代谢活动。细胞分化发生在多细胞生物体和单细胞生物体中。多细胞生物体中细胞的分化用来组成具有特定结构和功能的不同组织和器官，而单细胞生物体中细胞的分化则是为了适应环境条件变化所采用的不同生活方式。在这里只研究多细胞生物体中的细胞分化。

5.2 基因组作为体细胞的身份标记

生物体中不同类型的体细胞，如血细胞和神经细胞，在形态和表型上都有很大的差异。如何对这些不同类型的体细胞进行分类？考虑到同一个生物体中的所有不同类型的体细胞都是从同一个受精卵通过细胞分裂和细胞分化的方式产生的，它们的基因组基本相同。生物体中不同类型的体细胞在细胞分化过程中的基因表达谱不同。相比之下，不同个体的体细胞有不同的基因组。例如，人类染色体在个体间的差异，平均每1000个碱基对中就有一个不同（多态性）。据估计，人类基因组在大约10^6个位点上有不同版本（Wolpert，2009）。因此，基因组可以用来鉴别个体有机体中之间的体细胞。也就是说，一个个体的所有体细胞可以看作是同一实体具有的不同构象或所处的不同状态，而不同个体中的体细胞是不同的实体。

上述观点得到了以下事实的支持：免疫系统拒绝从其他人甚至近亲那里提取的干细胞，因为这些干细胞被受体视为外来干细胞（Wolpert，2009）。

基因组是体细胞这一个实体固有的身份标识。一组特定的基因（即奢侈基因）表达谱代表细胞的特定表型或特定状态。分化过程可以看作是分裂的细胞从一种状态转变为另一种状态的过程。例如，个体发育是分裂的细胞将其构象从全能状态转变为终末分化状态的过程。

5.3　体细胞的分化潜能和分化势能

多细胞生物体通常由不同等级层次的细胞组成，如干细胞、祖细胞和功能细胞。不同层次的细胞具有不同的分化能力。例如，受精卵的能力最高，而功能细胞的能力最低。此外，干细胞被分为全能干细胞（受精卵）、多潜能干细胞（如 ESCs）和专能干细胞（如 HSCs）等。为了更好地描述细胞的这种特性，需要引入一个新的概念，我把它称之为分化潜能。显然，一个细胞能分化的细胞种类越多，它的分化潜能就越高。因此，分化潜能可以定义为细胞能够分化成不同种类细胞的数目。如果一个细胞可以分化为 N 种不同种类的细胞，那么它的分化潜能被定义为 N。

如果我们更详细地研究这一过程，分化可以进一步分为直接分化和间接分化。直接分化是指细胞直接分化为另一种不同种类的细胞。间接分化是指分化的细胞进一步分化为另一种不同种类的细胞。因此，一个细胞可以分化的细胞类型的数目被定义为直接分化细胞种类的数目以及各层次中所有间接分化细胞种类数目的总和。图 5.1 给出了一个简单的示例。如图所示，A 细胞分化为 B_1 和 B_2 两种不同种类的细胞，这是属于直接分化。B_1 分化为 C_1、C_2、和 C_3，而 B_2 分化为 C_4 和 C_5。我们假定 C_1、C_2、C_3、C_4、C_5 是终末分化细胞。B_1 分化为 C_1、C_2 和 C_3 属于 B_1 的直接分化，B_2 分化为 C_4 和 C_5 属于 B_2 的直接分化。所有这些分化均为 A 的间接分化。根据上述定义，A 的分化潜能是 7，B_1 是 3，B_2 是 2，所有的 C 均是 0。

在分化过程中，细胞的基因组并没有改变，但某些特定基因组群的表达却发生了变化。事实上，分化是基因组中不同特定基因组群表达的结果。基因组的不同表达是由于染色质结构的不同构象所致（Ho & Crabtree，2010；Orkin & Hochedlinger，2011）。例如，实验结果表明，干细胞中的染色质具有"开放"构象，而体细胞中的染色质具有"关闭"构象（Meshorer & Mattout，2010；Gaspar–Maia，et al.，2011）。

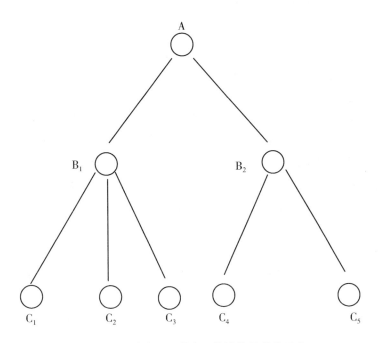

图 5.1　细胞在不同状态下的潜能的分化示意

为了更好地描述细胞在不同状态之间的转变，需要引入另一个新概念，我把它称之为分化势能（以下简称"势能"）。这个势能是分化状态的一个特征，并且可能与核内染色质的构象有关。两个状态之间的势能差异是两个状态之间如何发生转变的一个指标。如果一个细胞能自发地从一个状态转换到另一个状态，这就意味着初态的势能高于终态的势能。例如，在个体发育过程中，受精卵分裂并分化为终末分化的功能细胞。这一结果表明受精卵的势能高于功能细胞的势能。

然而，细胞并不总是能够自发地从较高势能的状态转换到任何其他较低势能的状态。更具体地说，不同的势能状态与染色质的不同构象有关。不同势能状态之间的转换对应于染色质构象的变化。如果构象的改变被某些因素所禁止，比如说拓扑方面的限制，那么即使初态比终态具有更高的势能，这个转换也是不允许的。

如果细胞想要从较低的势能状态转换（或跳跃）到较高的势能状态，该细胞必须克服这个势垒，这只能通过外部的帮助来完成。例如，体细胞可以通过在培养中引入四种转录因子，从终末分化状态转化为多潜能干细胞状态（Takahashi & Yamanaka，2006）。

染色质在分化过程中构象变化的机制是什么？换句话说，在细胞分化过程中，染色质一系列不同构象之间有什么关系？

现在还很难详细回答这个问题。然而，下面的类比也许会为理解这种关系的要点提供一些线索。考虑弹簧这个简谐动力学系统。受压弹簧的松弛过程是一个自发的过程，而压缩弹簧需要外力。类似于这个简谐动力学系统，我们可以假设，在分化过程中，染色质在一系列不同的势能状态中具有拓扑同源的构型，"压缩"较大的构型对应于势能较高的状态，而"压缩"较小的构型则对应于势能较低的状态。因此，分化过程是染色质从"压缩"较大的构型转变为"压缩"较小的构型的过程。当然，细胞在分化过程中不同势能状态之间的具体转换机制还需要进一步的研究。

分化潜能和分化势能是两个相关但又不同的概念。分化潜能表示一个细胞能分化出多少种细胞类型，而分化势能是与染色质的构象相关的特征量，其中两个状态的分化势能之差则显示细胞在两种不同状态之间转变的可能途径。

5.4 细胞分化假设

在多细胞生物体中，细胞分化产生不同类型的细胞，从而发挥不同的功能。例如，在脊椎动物（包括哺乳动物）和人类中，受精卵分化产生出200多种不同类型的细胞。细胞分化是自发过程，因此其分化能力应该是细胞的内禀属性。那么，这个过程的驱动力是什么呢？根据前面几节的讨论，我提出一个假设来解释这种驱动力的机制。

细胞分化假设：体细胞具有使其分化潜能最小化的趋势。

这一假设表明细胞具有自发地降低其分化潜能的内禀特性。由于细胞分化是一个与环境密切相关的过程，所以我们在某些过程中可能观察到与假设所预测的不同行为。这些现象可能是由于与邻近细胞的相互作用或来自环境的干预，从而抑制了细胞的这种内禀特性。

一般来说，分化途径中不同类型的细胞具有不同的分化潜能。例如，胚胎干细胞和造血干细胞具有不同的分化潜能。根据假设，所有的细胞都有降低其分化潜能的趋势。换句话说，如果想要阻止细胞分化，则必须提供一些特殊的微环境，比如干细胞壁龛，以保持细胞处于相同的分化潜能状态。因此，除了处于终末分化状态的细胞，每个状态下的细胞都需要适当的微环境来维持其分化状态。根据实验观察，可以推断这种微环境容纳细胞的能力是非常有限的。例如，哺乳动物在八细胞期的桑葚胚细胞，具有与受精卵相同的分化潜能，在一定条件下可以分化和发育成一个完整的个体。换句话说，哺乳动物的微环境，或干细胞壁龛，只能容纳 8 个全能细胞。如果已经有 8 个细胞，而且细胞还在继续分裂，那么干细胞壁龛就没有足够的能力来容纳额外的细胞。在这种情况下，额外的细胞会分化成具有分化能力较低的细胞。这样，就形成了细胞分化潜能的等级层次。这个细胞的层次结构也许类似于具有电子能级的原子结构。

人类（及其他哺乳动物）的个体发育过程可以用有丝分裂和细胞生长假设和细胞分化假设来解释。实验结果表明，多细胞生物体中的大多数组织都有相应的干细胞，且干细胞数目很小。可以推断，组织中的干细胞壁龛只能容纳为数不多的干细胞。当受精卵分裂时，分裂的细胞将首先填充全能干细胞壁龛。这些分裂的细胞继续分裂，由此产生的细胞将填满所有全能干细胞壁龛。那么再分裂出来的细胞将分化到下一级较低的分化潜能状态。现在处于这一潜能级状态的细胞有两个不同的来源：（1）从上一潜能级的细胞分化而来；（2）从同一潜能级的细胞分裂而来。似乎合理的假设是：第二个来源提供的细胞多于第一个来源提供的细胞，因为第二个来源的细胞只经历一次分裂过程，而第一个来源的细胞必须经历两个过程：一次分裂和一次分化。当这一潜能级的所有状态都被占据时，上一级全能状态的细胞将会停止分裂。其原因是这时分裂的细胞不能分化到下一潜能级的状态，而根据细胞分裂的经济原则，在这种情况下细胞不会分裂。在同一潜能级进一步分裂的细胞将分化到下一潜能级的细胞。

这样的过程可以继续重复下去。因此，在生物体中会有一系列的具有不同潜能的干细胞和祖细胞。祖细胞进一步分化为终末分化的功能细胞。这些功能细胞不能再分化了，因为它们的分化潜能是零。原则上，当终末分化状态被完全填满时，分化途径上的整个细胞系列将停止分裂，个体发育过程也就完成了。然而，在实际上，这种情况可能很少发生，因为生物体在本质上是动态的，细胞在不断变化。可以推测，生物体中细胞类型的数目以及每种类型细胞的数量的信息都应该包含在细胞的基因组中。

在胚胎发生过程中，细胞的分化潜能逐渐从全能性向多能性转变，最终向单能性分化。例如，两栖动物囊胚形成前的卵裂球具有与受精卵相同的分化潜能。这些细胞具有全能性，能够在一定条件下分化和发育成完整的个体。在形成三个胚层后，由于细胞的位置和微环境的差异，细胞的分化潜能仅限于分化为相应生殖层特定组织和器官的细胞。这些细胞是多能性的。器官形成后，组织中所有细胞的命运已经决定了，即它们只能分化为特定类型的细胞。现在这些细胞是单能性的。

随着受精卵分化潜能的降低，细胞的染色质结构也相应地变化。在这个过程中，不同的基因群组被表达。那么细胞核在这个过程中是否有任何变化？核移植技术的实验结果表明，分化体细胞的细胞核保留了发育正常个体所需的整套基因，细胞仍然具有发育成整个生物体的能力。而且，体细胞细胞核内的染色质结构在卵细胞的细胞质中可以恢复到全能性的结构。也就是说，受精卵在分化过程中染色质结构的变化是可以恢复的。从这个意义上说，受精卵的分化过程对染色质结构是可逆的，即在某些条件下，终末分化细胞的染色质结构可以恢复到与受精卵具有相同功能的结构。

此外，实验结果表明，通过添加一些适当的转录因子，终末分化的细胞可以恢复到胚胎干细胞的状态。例如，在逆转录病毒载体的帮助下，将四种转录因子（OCT 3/4、Sox 2、c–Myc、Klf4）的基因注入小鼠皮肤成纤维细胞中，成年小鼠成纤维细胞重新获得了胚胎干细胞的多能性（Takahashi & Yamanaka, 2006）。

5.5　分化性与功能性的关系

在细胞分化过程中，功能与分化程度密切相关。细胞分化越多，其功能就越确定。为了进一步研究它们之间的关系，需要引入细胞的功能性和分化性的概念。功能性可以定义为细胞在实际生化过程中履行其功能能力的百分比。终末分化细胞的功能是完全确定的，因此其功能性是最高的，可以取为 1；而全能干细胞的功能则完全不确定。也就是说，尚未向任何功能细胞分化，因此其功能性最低，可取为 0。分化性可以定义为细胞在特定分化途径上分化为所有不同类型细胞的能力的百分比。假设一个全能干细胞经历 N 种不同类型的细胞，成为一个终末分化细胞。如果一个细胞处于一种状态，它可以分化成 n 种不同类型的细胞，沿着同一分化途径到达终末分化状态，那么细胞在该状态下的分化性定义为 n/N。显然，全能干细胞的分化性为 1，而终末分化细胞的分化性为 0。

对实验结果的分析表明，细胞的分化性与功能性之间可能存在一定的定量关系。当细胞的功能性很高时，其分化性就很低。相反，当细胞的功能性很低时，其分化性就很高。例如，全能干细胞的分化性最高，其功能性最低。相比之下，终末分化细胞的分化性最低，其功能性最高。在自发分化过程中，细胞的分化性降低，其功能性增强。在诱导多能干细胞的过程中，细胞的分化性增强，其功能性减弱。如果 F = 功能性，D = 分化性，那么在分化和去分化的过程中，下列公式应该成立：

$$F + D = 1 \qquad (5-1)$$

因此，在细胞分化和去分化过程中，细胞的分化性和功能性之和是守恒的。

细胞的分化性与功能性之间的关系可以用分化性 – 功能性图来描述，如图 5.2 所示。细胞的状态可以用图中的一个点来表示，而细胞在分化或去分化过程中的所有状态可以用一条线来描述。例如，全能状态可以用点（1，0）表示，其中第一个数字表示分化性，第二个数字表示功能性；终末分化状态可以用点（0，1）表示。显然，所有其他状态都应该用这两个状态之间的线上的点来表示。例如，在分化过程中，如果一个细胞处于这样的状态，它通过分化 N/2 种不同的细胞类型达到终末分化状态，那么它的分化率为 0.5（其中 N 是一个全能细胞达到最终分化状态的不同细胞类型的数目）。在这个状态下，可以合理地认为细胞的功能性也是 0.5。因此，分化性 – 功能性图清楚地显示了分化过程中细胞状态的变化。个体发育过程由一条从（1，0）点到（0，1）点向下移动的直线来表示，而去分化过程则用向上移动的直线来表示。相比之下，细胞分裂仅用图表中的一点来表示，因为在分裂过程中，无论是分化性还是功能性都没有变化。

物理科学中的"基态"是指系统的最低能量状态或最稳定状态。如何定义生命系统的"基态"？应该用什么量来确定"基态"？从上述讨论可知，分化潜能可能是这一目的的合适选择。例如，处于全能状态的细胞在个体发育过程中会自发地不断地分裂和分化。在此过程中，分化潜能不断下降。当细胞到达终末分化的功能状态时，细胞不能进一步分化，分化潜能也不能进一步降低。这种情况类似于重力势能，当地球上的物体从高处往低处下降时，重力势能就会减小。因此，终末分化状态可以看作是细胞在分化过程中的"基态"，在分化性 – 功能性图中用（0，1）表示。

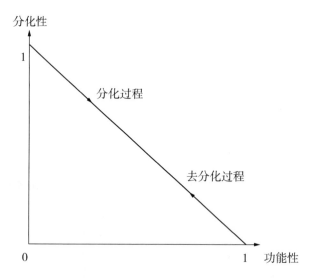

图 5.2　细胞在分化和去分化过程中的分化性和功能性的变化

5.6　多细胞生物中体细胞的树状结构

正如 5.2 节所讨论的那样，一个多细胞生物体中的所有体细胞都可以被看作是同一个实体在不同构象或不同状态的表现。这些细胞的等级关系可根据其分化潜能进行分类。追溯生物体的发育过程，我们可以构造一个图表来显示一个生物体中细胞的等级关系，这可以被称为生物体中细胞的树状结构。图 5.3 显示了这样一个结构的示意图。如图所示，树的生长始于原始细胞，如哺乳动物的受精卵或桑葚细胞。这些细胞分裂并分化为多潜能胚胎干细胞。然后，树分支到不同类型的专能成体干细胞。如 5.4 节所述，成体干细胞可以继续分裂和分化，直到最终分化成功能细胞为止。在这个结构中，具有全能的细胞被认为是处于根状态，而功能细胞被认为是处于叶状态。这两种状态之间的细胞被认为是处于茎状态，包括从多潜能干细胞到单能干细胞。根据从全能根状态分化到茎状态的次数，可将茎状态进一步划分为一级茎状态、二级茎状态和三级茎状态。因此，树状结构清楚地显示了完全基于分化潜能细胞在生物体层次结构中的位置。

树状结构为细胞在生物体中的功能提供了清晰的图像。处于叶状态的终末分化细胞参与生物体的所有具体和实际的生物学功能，如神经细胞和

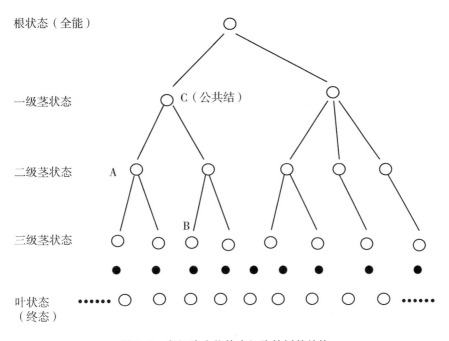

图5.3 多细胞生物体中细胞的树状结构

血细胞；而处于茎状态的细胞则在生物体中充当管家，例如替代受损、老化或死亡的细胞。在树状结构的基础上，可以提出一种修复损伤、老化或死亡细胞的等级机制。当处于叶状态的细胞受损、老化或死亡而需要更换时，处于同一叶状态的细胞如果能够完成任务，就会分裂以取代它们。如果不能的话，最靠近叶状态的干细胞将进行分裂和分化以取代它们。同样，如果需要更换处于茎状态的一些细胞，而处于同一茎状态的其他细胞不能满足这一要求，则处于茎状态最近的上层细胞将分裂并分化以完成这项工作。显然，这种等级修复机制使生物体中的资源得到了最佳利用，并最有效地维持了生物体。这种机制得到以下事实的支持：造血干细胞等干细胞是处于休眠状态的。据估计，这些干细胞小鼠一生只分裂5次，而大多数种群再生活动是由在干细胞分化级联下游的已定向的祖细胞完成（Zipori，2009）。

5.7　不同分化潜能态之间的转换

细胞能够在分化、去分化和转分化的过程中改变其类型。例如，在胚胎发生过程中，受精卵通过细胞分裂和分化从全能状态转变为功能状态。另一方面，实验结果表明，通过重新编程（Hanna, et al., 2008）、体细胞核移植（SCNT）(Jaenisch & Young, 2008)、加入一些"核心"转录因子如 Oct‐4 和 Sox‐2（Takahashi & Yamanaka, 2006）或小分子化合物（Hou, et al., 2013），可提高分化功能细胞的分化潜能。这些结果清楚地证明了细胞在分化或去分化过程中可以改变其分化潜能。细胞在分化潜能态的这种变化可以看作是细胞在这些状态之间的转换。

在转分化过程中，细胞可以从一种类型转变为不同分化途径的另一种类型的细胞。根据树状结构，可以提出一种可能的转分化机制。如果细胞想要从一种类型转变成在不同分化途径中的另一种不同细胞，它必须首先去分化到至少最接近的更高潜能的状态，这个状态对于初态和终态的分化途径都是共同的，称为公共结，如图 5.3 所示。然后再从这个公共结分化到终态。例如，如果处于 A 状态的细胞要向 B 状态转换，则细胞需要从 A 状态去分化到 C 状态，然后再从 C 状态重分化到 B 状态（如图 5.3 所示）。转分化过程的真正机制需要进一步的研究来揭示。

生命本身就是一种计算现象。

——克劳斯·埃米切（Claus Emmeche）

第六章 理论生物学的"终极法则"猜想

6.1 背景知识

生命是自然界一种神奇的现象。它无处不在地存在于我们这个物质世界中，给我们带来了丰富多彩的景色。生物学研究的主要目的之一就是理解生命的本质和内涵。在人类历史上，我们从亚里士多德时代就开始了探索生命意义的旅程，并坚持不懈地追求生命的真理。我们付出了很大的努力并已经取得了一些成绩，但我们仍然未能完全理解生命的本质和内涵。当科学取得突破时，我们通常期望这些科学突破能够帮助我们更好地理解生命的本质和意义。然而，结果并不总是令人满意。造成这种情况的原因有很多，其中一些原因可能是我们知识和技术的局限性。

正如理查德·费曼（Richard P. Feynman）所说，"我不能创造的事物，我就不能理解"。为了理解生命，科学家们一直在努力合成人造生命，希望由此来揭示生命的机制和原理。而人造生命就是由纯化学成分合成并能够进行达尔文进化过程的系统。史蒂文·本纳（Steven Benner）指出，我们可以把是否能够合成这样一个系统作为衡量一种生命理论正确与否的标准。他生动地描述："如果生命只不过是一个能够进行达尔文式进化和自我维持的化学系统，而且我们真正理解化学支持进化的机理，那么我们就应该能够合成一个能够进行达尔文式进化的人造化学系统。如果我们成功了，那么支持我们成功的理论将被证明是赋予我们威力的……相反，如果我们努力创造一个化学系统之后还不能得到一个人造生命形式……我们必须得承认，我们的生命理论还有欠缺。"（Benner，2009）

生命最重要的特征之一是自我复制，这无疑是合成人工生命的首要目标。科学发展的历史表明，合成生命的工作始于计算机。20世纪30年代，理论计算机科学和人工智能之父艾伦·图灵（Alan Turing）发明了一种通用计算机,称为通用图灵机。原则上,这台机器可以用来计算写在磁带上的任何可计算的序列。在20世纪40年代,约翰·冯·诺依曼(John von Neumann)发展了图灵的思想,并建造了一台自我复制机器。通过一系列的逻辑操作,机器可以创建一个自己的副本,而这个副本又可以产生另一个自己的副本。自从1953年DNA分子双螺旋结构的确定,DNA在遗传学中的生物学地位就建立起来了。分子生物学的发展在分子水平上增进了我们对生命系统基本过程的理解。DNA是遗传信息的载体,细胞基因组包含了细胞的所有信息。因此,DNA序列对于我们理解生命的本质是非常重要的。

近几十年来,基因组测序技术得到了突飞猛进的发展。今天,包括人类在内的许多物种的基因组序列已经确定。随着这项技术的改进,我们将有可能对我们感兴趣的任何物种的基因组进行测序。我们可能会问的最重要的问题之一是,我们能够从基因组序列中获得多少信息以及这些信息有多可靠。换句话说,我们如何从现有的基因组序列中理解生命的本质?这确实是我们需要回答的一个关键问题。在本章中,我将分析计算机编程语言的机器码与活细胞基因组之间的相似性,并试图回答从细胞基因组中能够提取哪些信息的问题。

6.2 人工智能

最近人工智能(AI)再次引起了人类的广泛关注。2016年3月,AlphaGo,一个围棋游戏的电脑程序,在五番棋比赛中以4比1的最终比分击败了世界冠军李世石。这是人类历史上第一次在围棋比赛中计算机程序击败世界职业冠军。围棋通常被认为是人类最复杂的经典游戏,也是计算机最难征服我们人类的经典游戏。但AlphaGo现在成功了!AlphaGo的设计与以往围棋程序的设计不同,它使用两个不同的神经网络进行博弈,一个神经网络用"价值网络"来估计形势,另一个使用"策略网络"来选择行棋(Silver, et al., 2016)。AlphaGo所采用的"深度学习"技术的主要原理类似于生物神经大脑的工作原理:形成"脑"神经网络进行精确计算,通过实例和经验进一步加强该网络。

发展人工智能的主要目标之一是使机器能够像人类那样"思考"。机器思考实际上是计算机的计算。一般来说,计算机主要做三件事:接收输入、进行计算和发送输出。不管输入和输出的内容是什么,计算机只把它

第六章 理论生物学的"终极法则"猜想

们当作一堆数据。为了使机器能够通过计算机程序进行思考，科学家们一直在研究智能算法，它可以根据输入数据计算所有可能性，从而正确有效地获得实际情况下的优化结果。从理论上讲，利用人工智能算法，机器可以完成人类所做的任何重复性工作，甚至是人类无法完成的工作。

为了使人工智能能够模拟人类的思维过程，科学家和工程师们对人类思维过程进行了研究。人脑是由神经元组成的复杂神经网络。每个神经元看起来都很简单，只是接收一些神经元的电信号，然后发出电信号来刺激其他神经元。虽然神经元的工作机制非常简单，但只要神经元数量足够多从而形成相互连接的神经网络，就可以执行其智能功能，例如，人类大脑能够进行智能性的工作，因为它包含大约 1000 亿个神经元，平均每个神经元大约有 7000 个突触与其他神经元连接。

人工神经网络的研究有着悠久的历史，可以追溯到 1943 年，当时麦克卡罗奇（W. S. McCulloch）和皮茨（W. Pitts）建立了神经元的数学模型，称为 MP 模型。借助该模型，他们证明单个神经元可以实现逻辑功能，从而开启了人工神经网络研究的时代。20 世纪 60 年代，Widrow 提出了一种自适应线性网络，在此基础上非线性多层自适应网络得到了发展。80 年代，霍普菲尔德（Hopfield）建立了神经网络模型，Linsker 提出了感知机网络自组织的新理论。近年来，人工神经网络主要朝着模拟人类认知方向发展，将模糊系统、遗传算法和进化机制相结合，发展成为计算智能。而且，作为神经科学中一个重要概念的卷积被引入到人工神经网络中，已经在语音识别和图像分类等领域取得了巨大的成果。

人工神经网络是一种用于信息处理的数学模型，它类似于人脑中突触连接的结构。它由大量的节点（或神经元）组成，可以接收上层"神经元"的输入信号，并根据"神经元"的重要性分配不同的权重。然后，"神经元"将加权输入信号进行汇总，并将结果集成到一个函数中。经过计算，最终结果将输出到神经网络下一层的"神经元"。人工神经网络通过一系列复杂的算法和大量的数据训练，可以像人脑中的神经网络那样工作。它可以从复杂的数据中找出"特征"，并产生需要智能思考的结果。

那么人工神经网络是如何学习的呢？人工神经网络的学习过程实质上是调整每个神经元的权重，使整个人工神经网络能够很好地满足测试任务的要求。以分类任务为例。首先，将学习集中的每个元素输入到人工神经网络中，并对网络进行分类训练。在完成整个学习集之后，网络将根据它从学习到的例子中获得的经验总结自己的观点和规律。然后，我们将用测试集中的例子来测试网络。如果网络通过了测试（例如90%的正确率），那么它已经被组建成功，可以用来处理分类事务。

人工神经网络具有自适应性和自组织性。它可以改变学习或训练过程中的突触权重，以适应环境的要求。它也是一个具有学习能力的系统，它将丰富自己的知识，使它能够拥有比设计者最初输入的知识还要多的知识。通常有两种训练方式。一种是在监督下的学习，对给定的标准样本进行分类或模仿。另一种是没有监督的学习，其中只提供了学习的方式或规则。学习的具体内容随环境（即输入信号）的不同而变化，系统可以自动发现其特征和规律。这种训练方式更类似于人类的训练方式。例如，围棋游戏的程序采用了两种不同的训练方法。AlphaGo 接受监督训练，AlphaGo Zero 则接受没有监督训练。AlphaGo Zero 可以完全依靠自己的学习能力达到专业棋手的水平，而不需任何研究人员或输入数据的帮助。结果表明，AlphaGo Zero 的性能明显优于 AlphaGo（Silver, et al., 2017）。因此，人工智能现在能够以人类的方式来"学习"和"思考"了。

在 AlphaGo 和 AlphaGo Zero 中使用的思想和技术也可以应用到需要长期规划和决策的其他人工智能领域，例如蛋白质折叠、减少能量消耗和寻找新材料的问题。这些成就表明了人工智能的强大未来。然而，原则上，不管人工智能是多么聪明，它的所有功能和活动都是由人类设计和完成的计算机程序来确定的。这些程序首先是用一些高级编程语言编写而成，诸如 C++、Java、Python、Prolog 和 Lisp。这种形式被称为程序的源代码。然后这些源代码被编译成由基本元素 0 和 1 不同组合组成的长字符串，它们被称为程序的机器代码。人工智能可以根据从这些机器代码中获得的信息和指令来行动。显然，不同设计水平的源代码会导致人工智能不同水平的行为，如 20 世纪 80 年代的简单机械动作和近年来的 AlphaGo 和 AlphaGo Zero 复杂神经网络"思维"。因此，算法的根本变化将改变人工智能的"思维"方式和行为特征。换句话说，计算机技术和算法的革命可以完全改变人工智能的智能水平。该过程可被称为人工智能的演化。AlphaGo 和 AlphaGo Zero 等人工智能的优秀表现令我们相信，人工神经网络的简单机制能够产生像人类一样复杂的行为以及精细的思维。

6.3 活细胞基因组的一些特征

细胞是所有生物体中最小的结构单位和功能单位。生命系统中大部分生物过程都是在细胞内进行的。正如威尔逊（E. B. Wilson）所说，所有生物学的答案最终都要到细胞中寻找，因为所有生命体都是，或者曾经是一个细胞。生物学的根本问题最终都是细胞的问题。从结构上看，真核细胞由细胞膜、细胞质和细胞核组成，而染色体 DNA 储存在细胞核内。这三个组成部分中的每一部分都在细胞中发挥作用，它们相互配合以维持细胞的正常功能。相比之下，原核细胞没有细胞核，染色体 DNA 储存在类核区。

薛定谔是第一个从物理科学的角度来研究生物学基本概念和基本原理的人。他认为生物系统的遗传行为是由染色体纤维这种"非周期固体"所决定的（Schrödinger，1944）。詹姆斯·沃森（James Watson）和弗朗西斯·克里克（Francis Crick）基于罗莎琳德·富兰克林（Rosalind Franklin）研究得出的 X 射线衍射图提出了 DNA 分子的双螺旋模型，显示出 DNA 在遗传学中的生物学作用（Watson & Crick，1953）。分子生物学的进一步发展提高了我们对基因在生物系统中的重要作用的认识。

普遍认为，根据从基因组中获得的信息，细胞可以按照正确的顺序在正确的时间执行所有的功能、完成所有的任务。科学研究表明，基因组包含细胞的所有信息。更具体地说，在分子水平上，中心法则表明遗传信息从 DNA 流向 RNA 到蛋白质。几乎所有细胞的特点和特征都是由蛋白质来表现的，细胞的功能主要是由蛋白质来执行的。基因组决定了哪些蛋白质被合成，在什么时候这些蛋白质被合成和调控。在细胞水平上，基因组在细胞生长、细胞分裂和细胞分化等生物过程中指导着细胞的行为。

我们对细胞基因组行为的理解已经取得了很大进展。发育生物学研究表明，卵母细胞的细胞质与基因组产生直接或间接相互作用，从而改变基因的表达模式，称为细胞质记忆。例如，将培养的爪蟾肾细胞核注射到蝾螈的去核卵母细胞中，结果表明，该细胞没有合成通常在肾细胞内合成的蛋白质，而只合成通常在卵母细胞中合成的蛋白质。因此，一个卵母细胞的细胞质能够激活体细胞内一些原本没有活性的基因，也能使一些原本有活性的基因失活，这显示出基因组的构象受到环境的影响。

利用体细胞核移植（SCNT）技术进行的系统研究表明，卵母细胞的细胞质可以改变基因组的表达模式。体细胞核移植是将卵母细胞的细胞核替换为体细胞（如皮肤细胞或肝细胞）的细胞核的过程。去核卵母细胞与体细胞核通过将体细胞核插入"空"卵母细胞而融合。寄主卵母细胞可以对体细胞核进行再编程，这样组合而成的细胞表现为胚胎干细胞，其活性和功能主要由供体体细胞的基因组所支配（Wilmut, et al., 2002）。这些结果表明，从受精卵分化而来的体细胞核保留了生物体发育所需的一切信息。换句话说，在某种意义上，染色质在分化过程中的结构变化是可逆的：只要条件合适，体细胞中的染色质结构就可以恢复到胚胎干细胞的状态。例如，伊恩·威尔穆特（Ian Wilmut）领导的一个研究小组采用核移植的方法成功地从培养的细胞系中成功地克隆了绵羊"多莉"。

克雷格·文特尔（J. Craig Venter）和他的研究团队最近的工作清楚地证明了 DNA 是生命的软件。文特尔和他的同事化学合成了丝状支原体的整个基因组，并将它移植到山羊支原体细胞中（Lartigue, et al., 2007）。他们发现合成基因组控制并取代了野生型山羊支原体细胞的原始基因组，而且受体细胞变成了丝状支原体细胞。这是人类历史上第一次将完全由化学合成的基因组（裸 DNA）移植到细菌细胞中并产生活细胞。他们的研究表明，完全由化学合成的基因组可以像活细胞中的天然基因组一样工作，也就是说活细胞有可能由计算机的数字信息所控制。文特尔在他的书中总结道："至少大多数分子生物学家认为，由计算机中的字母序列所代表的 DNA 和基因组是生命的信息系统。现在我们从计算机中的数字信息开始，实现了这个回路的闭合，并仅通过使用这些信息，化学合成和组装了一个完整的细菌基因组，并将其移植到受体细胞中，从而产生了一个仅由合成基因组控制的新细胞。"（Venter, 2013）最近，一种数字到生物的转换器被设计完成，它可以从计算机接收数字信息并转化为生物聚合物，如 DNA、RNA 和蛋白质（Boles, et al., 2017）。这标志着向合成人造生命迈出了一大步。

上述讨论清楚地说明了 DNA 在活细胞中的重要作用。如果我们替换了一个生物体的基因组，我们就替换了生物体的物种。虽然上述结果只是替换生命，而不是严格意义上的创造生命，但意义仍然是深远的。生命的特征，包括物种，是由基因组序列独自决定的。

6.4 理论生物学的"终极法则"

在前面的 6.2 和 6.3 节中,我们讨论了人工智能的能力以及活细胞基因组的特征。计算机科学技术的进步是惊人的,表现在人工智能从简单的机械动作演化到像人类那样的复杂逻辑思维和自我学习。另一方面,人类行为的复杂而精致的方式可能来源于由 A、G、C 和 T 这四个核苷酸组成的长串组合的基因组中所包含的信息。

的确,注意到细胞的基因组和人工智能的机器代码之间有着惊人的相似性是非常重要的。基因组有 A、G、C 和 T 四个元素,而机器代码有 0 和 1 两个元素。我们相信基因组包含了细胞的所有信息而机器代码包含了人工智能的所有信息。人工智能根据从机器代码中获得的指示和指令来响应刺激,而细胞的情形可能与人工智能类似。

根据物理学的观点,薛定谔推测染色体必须包含"某种决定个体未来发展的整个模式的密码本",他推断,密码本必须包含"有序的原子组合,具有足够的抵抗力来永久地保持其有序性"。他还解释了"非周期性固体"中的原子数如何携带足够的遗传信息。他论证说这种固体不一定要非常复杂才能容纳大量的排列,可能是像莫尔斯电码(Morse code)那样的一个简单的二进制密码(Schrödinger,1944)。

沃森和克里克发现了 DNA 的双螺旋结构,并展示了密码本将遗传信息代代相传的具体方式。在双螺旋结构下,DNA 分子由两条链组成,每条链都向相反的方向运动。一条链中的核苷酸与另一条链中的核苷酸互补:腺嘌呤(A)与胸腺嘧啶(T)的碱基配对,鸟嘌呤(G)与胞嘧啶(C)的碱基配对。因此,DNA 中的每一条链都可以作为模板来重建另一条链。这样,基因组中包含的信息就可以被复制并传递给后代。

20 世纪 70 年代,弗雷德里克·桑格(Frederick Sanger)提出了用链终止抑制剂进行 DNA 测序的方法,并确定了第一个全基因组的序列——噬菌体 φX174 基因组(Sanger, et al., 1977)。2001,第一个完整的人类基因组序列被确定,我们第一次真正看到了包含人类生命密码的非周期性固体的具体细节。

确实,我们在基因组测序方面取得了很大进展。尽管我们相信活细胞的基因组包含了细胞的所有信息,如细胞的生长、分裂和分化,但我们从序列本身只能获取非常有限的信息。编码区通常只占基因组的 2% 左右。

虽然基因组中的一些非编码区可能在调控基因表达、组织染色体结构和控制表观遗传等方面发挥一定的作用，但我们还是不了解基因组中许多区域的生物学功能和作用。

计算机科学家的任务是根据编程语言的语法（或规则）通过优化程序中的结构和策略来设计出最聪明、最有效的人工智能。由于人工智能不能理解这些高级语言，所以用编程语言编写的程序，或程序的源代码，必须编译成机器代码，以便人工智能能够理解程序中的指令并相应地执行功能（如图6.1A所示）。

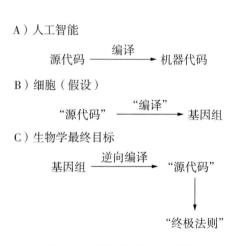

图6.1　"终极法则"的示意

类似于人工智能，细胞可能只能理解基因组，并根据基因组中包含的信息履行功能。因此，可以假设细胞的基因组是"机器代码"，它们是通过编译用"生物编程语言"为细胞编写的程序而得到（图6.1B）。这些编程语言的语法或规则可以被认为是理论生物学的"终极法则"（图6.1C）。在这里，我提出一个关于理论生物学的"终极规则"猜想：

"终极法则"猜想："终极法则"是"生物编程语言"的基本原则，它们可以通过对细胞基因组的逆向编译而获得。

根据这个猜想，生物科学家的最终目标是破译细胞的基因组并发现这些"终极法则"。事实上，分子生物学的研究在发现"终极法则"方面已经取得了一定的进展。例如，我们已经发现了一些生物编程语言的相关元素，如遗传密码中的"开始"和"停止"密码子，它们指示着蛋白质翻译过程的开始和结束。另外，近年来对非编码 DNA 序列的研究表明，这些区域在调控基因表达中起着重要的作用。表观遗传修饰提供了另一种方式来调节基因的表达。这些结果证明了 DNA 片段在调控细胞行为和特性中的作用。

　　当然，要实现这一目标还有很长的路要走。我们面前有许多障碍，我们仍然不清楚如何克服这些障碍。DNA 测序技术得到了极大的发展，其准确性也得到了很大的提高。现在我们有能力对任何我们感兴趣的基因组进行测序。然而，即使我们手上拥有所有基因组的序列，也不一定意味着我们理解生命的本质。一些科学家指出，我们拥有所有人类基因的序列，但我们无法理解它们。

　　如上所述，基因组可能只是"生物编程语言"的"机器代码"。如果这是真的，我们将很难从基因组序列中理解生命系统的行为和特征，因为我们从基因组序列中获得的信息可能是模棱两可甚至是误导的。因此，当我们解释基因组序列时，我们应该记住，基因组可能只是程序的"机器代码"。

　　道路是艰难的，但我仍然对未来感到乐观。尽管目前还不清楚我们到底该往哪里去，但目前的知识可能为将来寻找"终极法则"提供一些线索。在基因表达过程中，基因组序列中所包含的信息可以被揭示出来。实验结果表明，基因表达不仅与 DNA 片段的序列有关，而且与这些 DNA 片段在整个 DNA 分子的三维结构中的位置有关。在同一生物体不同类型的细胞中，基因表达的不同模式说明了这一点。更具体地说，我们都知道真核细胞的基因组是由细胞核内的组蛋白组装而成的，而在多细胞生物体中，具有相同基因组的细胞可能采用不同的细胞类型，从干细胞到不同的终末分化细胞。表明它们经历了不同的细胞分化途径，并具有不同的基因表达模式。可以认为，真核细胞的基因组不仅包含 DNA 序列中的信息，而且还包含 DNA 片段的空间排列信息。换句话说，相同 DNA 片段在不同空间排列可能代表不同的信息。显然，以这种方式存储的信息比仅存储在 DNA 序列中的信息要有效得多。因此，DNA 片段在真核生物基因组中的不同空间排列可能代表了生物编程语言的不同语句，也许类似于计算机编程语言中的"指针"。

　　相比之下，细菌基因组的组装方式与真核细胞的不同。在细菌中，遗传物质不是被膜包围在核内，而是可以在细胞质中与核糖体接触，因此

RNA 转录和蛋白质翻译过程可以同时发生。细菌的基因组本质上是"裸露"DNA，一种双链环状 DNA 分子，转录后的 mRNA 没有内含子，可以直接翻译成多肽。此外，DNA 复制发生在环形分子的两个方向。而且，细菌采用不同于真核生物的 DNA 甲基化策略来保护自己的 DNA 不被其限制性内切酶切割。例如，真核生物中克隆的一些基因在原核生物的表达系统无法识别。所有这些在基因组包装以及基因在原核细胞和真核细胞中具有表达的差异表明，这两种细胞的基因组可能是来自于编译生物编程语言的不同"算法"的"源代码"。

可以期待，分子生物学和生物科学的其他分支将极大地促进我们对基因组意义的理解及丰富我们对生物编程语言的知识。当我们获得"终极法则"时，我们将对生物世界的全貌有充分的了解。我们将知道生命系统中所有生物过程的基本机制，如人体内神秘的个体形成过程。例如，在个体形成过程中，身体不同部位的各种器官的形成也许取决于诸如"go to"和"proliferate when"这样的语句。推测所有这些信息也许都包含在先前认为是基因组的"垃圾 DNA"区域内。

生物学中的"终极法则"的地位可能与经典物理学中的牛顿定律相似。"终极法则"可以决定生命系统的行为和特征。可以预料，前几章所讨论的关于活细胞的生存公理和四个基本假设，全部都是属于活细胞的内禀性质，应该包含在"终极法则"中。此外，细胞与外界环境相互作用时应遵循的规律，这是生物学理论进一步研究的课题，也应包含在"终极法则"中。

6.5 "终极法则"的含意

当然，生命系统的基因组包含了所有的信息，这足以指导系统的行为和确定系统的特性。然而，我们如何才能正确地从基因组中挖掘出信息来理解控制生命系统的原理呢？从人工智能与细胞的比较中，我们知道基因组的 DNA 序列也许只是细胞的"机器代码"。因此，在基因组水平上理解生命系统的原理就类似于在机器代码水平上理解人工智能的原理。尽管人工智能的所有特性和行为都可以由高级编程语言编译而来的机器代码决定，但从机器代码（1 和 0 的长字符串）中很难理解人工智能的特点和原理。同样，虽然活细胞的基因组足以决定细胞的特性和行为，但从基因组，即生命的"机器代码"中理解细胞的原理也是非常困难的。因此，为了理解生命的原理和规律，我们需要破译"机器代码"来获得生物编程语言的"源代码"和"语法"。只有在"源代码"的层次上，我们才能完全理解生命系统和生命的本质。

诚然，"终极法则"的概念是初步的，现在更多的是概念性而非实用性。然而，它也许为解释某些理论预测与实验结果之间的不一致以及一些现有理论无法解释的观测提供了一种新的解释方法。此外，它还将为探索生命提供一个新的方向。

如6.1节所述，合成人造生命是理解生命本质的重要手段。这个宏大项目的第一步，也是最重要的一步，就是合成一个活的细胞，这将为我们理解生命打开一扇大门。丘奇和雷吉斯（Church and Regis）在他们的书中提出了合成细胞的定义："真正的合成细胞是我们从底层创造出来的细胞，这可能是一种由纯化学成分制成的新形式的生命物质"（Church & Regis, 2012）。文特尔和他的研究团队朝着这个目标迈出了巨大的一步。他们的结果表明，对于细菌来说完全由化学合成的基因组就像自然基因组一样完全控制着宿主细胞，基因组是生命的软件（Venter, 2013）。他们在实验中还发现，在多于一百万个碱基对中，一个碱基对的缺失会导致有生命和无生命的差别（Venter, 2013）。这一发现意味着很难在基因组水平上来改变生物体的特征。目前对有机体特性的改良研究基本上仍是靠碰运气。如果我们把基因组看成是生物体的"机器代码"，那么这种情况就可以解释。如果"终极法则"的概念是正确的，那么细胞的功能和作用就由用生物编程语言编写的程序来决定。我们很容易理解想通过改变程序的机器代码来修改人工智能的特性和行为是多么困难和不确定。同样，我们可以想象在"机器代码"水平上来理解生物程序是多么困难，而通过改变基因组中的核苷酸来改变生物体的特性是多么的不确定。

"终极法则"不仅可以帮助我们理解生命系统的基本过程，而且可以提供一种有效而直接的方法来修正生命系统的特性。近几十年来，合成生物学取得了很大的进步。通过重新设计和重新计算基因组，合成生物学家已经成功地将微生物改造得具备许多不同的特殊功能，例如生产燃料、检测饮用水中的砷（Church & Regis, 2012）。一旦获得了"终极法则"，我们将能够为合成生物学做出巨大贡献。"终极法则"将使生物合成更加容易和有效。例如，我们也许能实现丘奇和雷吉斯提出的建议"……使人类对所有已知或未知、自然或人为的病毒免疫"（Church & Regis, 2012）。此外，我们还可以使用生物编程语言编写一个程序来从头合成一个活细胞，这确实是一个真实意义上的合成细胞。我们期待"终极法则"将使我们对生物世界有一个全面和彻底的认识。

References

Allsopp, R. C. & Harley, C. B. 1995. Evidence for a critical telomere length in senescent human fibroblasts. *Experimental Cell Research*, 219, 130–136.

Barbieri, M. 2003. *The Organic Codes: An Introduction to Semantic Biology*. Cambridge: Cambridge University Press.

Benner, S. 2009. *Life, the Universe and the Scientific Method*. FL: Foundation for Applied Molecular Evolution.

Biessmann, H. & Mason, J. M. 2003. Telomerase-independent mechanisms of telomere elongation. *Cell. Mol. Life Sci.*, 60, 2325–2333.

Boles, K. S., Kannan, K., Gill, J., Felderman, M., Gouvis, H., Hubby, B., Kamrud, K. I., Venter, C., Gibson, D. G. 2017. Digital-to-biological converter for on-demand production of biologics. *Nature Biotechnology*, 35, 672–675.

Borh, N. 1933. Light and life. *Nature*, 131, 457–459.

Bryan, T. M. & Reddel, R. R. 1997. Telomere dynamics and telomerase activity in in vitro immortalised human cells. *Eur. J. Cancer*, 33, 767–773.

Campisi, J. 2000. Cancer, aging and cellular senescence. *In Vivo*, 14, 183–188.

Campisi, et al. 2013. Aging, cellular senescence, and cancer, *Annu. Rev. Physiol.*, 75, 685–705.

Church, G. & Regis, E. 2012. *Regenesis: How Synthetic Biology Will Reinvent Nature and Ourselves*. New York: Basic Books.

Counter, C. M., Meyerson, M., Eaton, E. N. & Weinberg, R. A. 1997. *Proc. Natl. Acad. Sci. U. S. A.*, 94, 9202–9207.

Cristofalo, et al. 2004. Replicative senescence: a critical review. *Mechanisms of Ageing and Development*, 125, 827–848.

Cristofalo, V. J., Volker, C., Francis, M. K. & Tresini, M. 1998. Age-dependent modifications of gene expression in human fibroblasts. *Critical Reviews in Eukaryotic Gene Expression*, 8, 43 – 80.

De Robertis, E. M. & Gurdon, J. P. 1977. Gene activation-in somatic nuclei after injection into amphibian oocytes. *Proc. Natl. Acad. Sci. U. S. A.*, 74, 2470 – 2474.

Durham, M. A., Neumann, A. A., Faschining, C. L. & Reddel, R. R. 2000. Telomere maintenance by recombination in human cells. *Nat. Genet.*, 26, 447 – 450.

Gaspar-Maia, A., Alajem, A., Meshorer, E. & Ramalho-Santos, M. 2011. Open chromatin in pluripotency and reprogramming. *Nature Reviews Mol. Cell Biol.* 12, 36 – 47.

Gershon, H. & Gershon, D. 2001. Critical assessment of paradigms in aging research. *Experimental Gerontology*, 36, 1035 – 1047.

Goldsmith, T. C. 2014. Modern evolutionary mechanics theories and resolving the programmed/non-programmed aging controversy. *Biochemistry (Moscow)*, 79, 1049 – 1055.

Goldsmith, T. C. 2015. Solving the programmed/non-programmed aging conundrum. *Curr. Aging Science*, 8, 34 – 40.

Green, D. R. 2011. *Means to an End: Apoptosis and Other Cell Death Mechanisms*. New York: Cold Spring Harbor Laboratory Press.

Greider, C. W. & Blackburn, E. H. 1989. A telomeric sequence in the RNA of Tetrahymena telomerase required for telomere repeat synthesis. *Nature*, 337, 331 – 337.

Hanna, J. et al. 2008. Direct reprogramming of terminally differentiated mature blymphocyte to pluripotency. *Cell*, 133, 250 – 264.

Harrison, D. E. 1985. Cell and tissue transplantation: a means of studying the aging process. *Handbook of the Biology of Aging*. Finch, C. E. & Schneider, E. L. (eds.), New York: Van Nostrand Reinhold Co.

Hawking, S. W. & Mlodinow, L. 2010. *The Grand Design*. New York: Bantam Books.

Higgins, D. A., Pomianek, M. E., Kaaml, C. M., Taylor, R. K.,

References

Semmelhack, M. F. & Bassler, B. L. 2007. The major vibrio cholerae autoinducer and its role in virulence factor production. *Nature*, 450, 883 – 886.

Ho, L. & Crabtree, G. R. 2010. Chromatin remodeling during development. *Nature*, 463, 474 – 484.

Hou, P. et al. 2013. Pluripotent stem cells induced from mouse somatic cells by small-molecule compounds. *Science*, 341, 651 – 654.

Jaenisch, R. & Young, R. 2008. Stem cells, the molecular circuitry of pluripotency and nuclear reprogramming. *Cell*, 132, 567 – 582.

Jammer, M. 1974. *The Philosophy of Quantum Mechanics*. New York: John Wiley & Sons, Inc.

Kauffman, S. 1995. *At Home in the Universe: The Search for the Laws of Self-Organization and Complexity*. New York: Oxford: Oxford University Press.

Klapper, W., Heidorn, K., Kuhne, K., Parwaresch, R. & Krupp, G. 1998a. Telomerase activity in "immortal" fish. *FEBS Letters*, 434, 409 – 412.

Klapper, W., Kuhne, K., Singh, K. K., Heidorn, K, Parwaresch, R. & Krupp, G. 1998b. Longevity lobsters is linked to ubiquitous telomerase expression. *FEBS Letters*, 439, 143 – 146.

Lartigue, C., Glass, J. I., Alperovich, N. et al. 2007. Genome transplantation in bacteria: Changing one species to another. *Science*, 317, 632 – 638.

Leonard, G. D., Fojo, T. & Bates, S. E. 2003. The role of ABC transporters in clinical practice. *Oncologist*, 8, 411 – 424.

Lingner, J. and Cech, T. R. 1996. Purification of telomerase from Euplotes aediculatus: Requirement of a primer 3' overhang. *Proc. Natl. Acad. Sci. U. S. A.*, 93, 10712 – 10717.

Mayr, E. 1961. Cause and effect in biology: Kinds of causes, predictability, and teleology are viewed by a practicing biologist. *Science*, 134, 1501 – 1506.

Mayr, E. 1988. *Toward a New Philosophy of Biology: Observations of an Evolutionist*. Cambridge: Harvard University Press.

Mayr, E. 1996. The autonomy of biology: the position of biology among the science. *The Quarterly Review of Biology*, 71, 97 – 106.

Mayr, E. 1997. *This Is Biology: The Science of the Living World*.

Cambridge: Harvard University Press.

Mayr, E. 2004. *What Makes Biology Unique?* Cambridge: Harvard University Press.

McClatchey, A. I. & Yap, A. S. 2012. Contact inhibition of proliferatin redux. *Curr. Opin. Cell Biol.*, 24, 685–694.

McShea, D. W. & Brandon, R. N. 2010. *Biology's First Law: The tendency for diversity & Complexity to increase in evolutionary systems.* Chicago: The University of Chicago Press.

Meshorer, E. & Mattout, A. 2010. Chromatin plasticity and genome organization in pluripotent embryonic stem cell. *Curr. Opin. Cell Biol.*, 22, 1–8.

Nadell, C. D., Xavier, J. B. & Foster, K. B. 2009. The sociobiology of biofilms. *FEMS Microbiol. Rev.*, 33, 206–224.

Nakamura, T. M., Morin, G. B., Chapman, K. B., Weinrich, S. L., Andrews, W. H., Lingner, J., Harley, C. B. and Cech, T. R. 1997. Telomerase catalytic subunit homologs from fission yeast and human. *Science*, 277, 955–959.

Nedelcu, A. M., Driscoll, W. W., Durand, P. M. & Herron, M. D. 2010. On the paradigm of altruistic suicide in the unicellular world. *Evolution*, 65, 3–20.

Orkin, S. H. & Hochedlinger, K. 2011. Chromatin connections to pluripotency and cellular reprogramming. *Cell*, 145, 835–850.

Prigogine, I. & Nicolis, G. 1971. Biological order, structure and instabilities. *Q. Rev. Biophys*, 4, 107–148.

Pross, A. 2012. *What Is Life? How Chemistry Becomes Biology.* Oxford: Oxford University Press.

Sanger, F. et al. 1977. Nucleotide sequence of bacteriophage phi X174 DNA. *Nature*, 265, 687–695.

Schmidt, E. V. 2004. *Coordination of Cell Growth and Cell Division. In Cell Growth: Control of Cell Size.* Ed. Hall, M. N., Raff, M. & Thomas, G. New York: Cold Spring Harbor Laboratory Press.

Schrödinger, E. 1944. *What Is Life? The Physical Aspect of the Living Cell.*

Cambridge: Cambridge University Press.

Sergiev, P. V., Dontsova, O. A. & Berezkin. 2015. Theories of aging: An ever-evolving field. *Acta Naturae*, 7, 9 – 18.

Silver, D. et al. 2016. Mastering the game of Go with deep neural networks and tree search. *Nature*, 529, 484 – 489.

Silver, D. et al. 2017. Mastering the game of Go without human knowledge. *Nature*, 550, 354 – 359.

Takahashi, K. & Yamanaka, S. 2006. Induction of pluripotent stem cells from mouse embryonic and adult fibrobalst cultures by defined factors. *Cell*, 126, 663 – 676.

Venter, J. C. 2013. *Life at the Speed of Light: From the Double Helix to the Dawn of Digital Life*. New York: Penguin Group (USA) LLC.

Watson, J. D. & Crick, F. H. 1953. A structure of deoxyribose nucleic acid. *Nature*, 171, 737 – 738.

Wilmut, I., Beaujean, N., de Sousa, P. A., Dinnyes, A., King, T. J., Paterson, L. A., Wells, D. N. & Young, L. E. 2002. Somatic cell nuclear transfer. *Nature*, 419, 583 – 586.

Wolpert, L. 2009. *How We Live &Why We Die: The Secret Lives of Cells*. New York: W. W. Norton & Company.

Xie, P. 2013. *Scaling Ecology to Understand Natural Design of Life Systems and Their Operations and Evolutions—Integration of Ecology, Genetics and Evolution through Reproduction*. Beijing: Science Press.

Zipori, D. 2009. *Biology of Stem Cells and Molecular Basis of the Stem State*. New York: Humana Press (Springer).